SpringerBriefs in Applied Sciences and Technology

SpringerBriefs present concise summaries of cutting-edge research and practical applications across a wide spectrum of fields. Featuring compact volumes of 50 to 125 pages, the series covers a range of content from professional to academic.

Typical publications can be:

- A timely report of state-of-the art methods
- An introduction to or a manual for the application of mathematical or computer techniques
- A bridge between new research results, as published in journal articles
- A snapshot of a hot or emerging topic
- An in-depth case study
- A presentation of core concepts that students must understand in order to make independent contributions

SpringerBriefs are characterized by fast, global electronic dissemination, standard publishing contracts, standardized manuscript preparation and formatting guidelines, and expedited production schedules.

On the one hand, **SpringerBriefs in Applied Sciences and Technology** are devoted to the publication of fundamentals and applications within the different classical engineering disciplines as well as in interdisciplinary fields that recently emerged between these areas. On the other hand, as the boundary separating fundamental research and applied technology is more and more dissolving, this series is particularly open to trans-disciplinary topics between fundamental science and engineering.

Indexed by EI-Compendex, SCOPUS and Springerlink.

Roberto Álvarez-Fernández ·
Alexandra Delgado-Jiménez ·
Fernando Beltrán-Cilleruelo

A Carbon Footprint Calculation Tool for Urban Development

 Springer

Roberto Álvarez-Fernández
Department of Engineering
Higher Polytechnic School
Nebrija University
Madrid, Spain

Alexandra Delgado-Jiménez
Department of Architecture
Higher Polytechnic School
Nebrija University
Madrid, Spain

Fernando Beltrán-Cilleruelo
Department of Engineering
Higher Polytechnic School
Nebrija University
Madrid, Spain

ISSN 2191-530X   ISSN 2191-5318   (electronic)
SpringerBriefs in Applied Sciences and Technology
ISBN 978-3-031-69891-0   ISBN 978-3-031-69892-7   (eBook)
https://doi.org/10.1007/978-3-031-69892-7

This Springer imprint is published by the registered company Springer Nature Switzerland AG
The registered company address is: Gewerbestrasse 11, 6330 Cham, Switzerland

If disposing of this product, please recycle the paper.

Introduction

This book is oriented to help developing a carbon footprint calculation tool for urban planning. The starting point is the evolution of urban planning related to environmental problems that has resulted in current situation. Now it is necessary to plan cities in a context of climate crisis. Planning with climate change criteria is especially important for urban growth or transformation control.

The framework was born in the Paris Agreement, it is a legally binding international treaty on climate change. It was adopted by 196 Parties at COP21 in Paris on 12 December 2015 and entered into force on 4 November 2016. It aims to limit global warming to well below 2, preferably 1.5 °C, compared to pre-industrial levels.

And it is necessary to create tools to measure the impact on climate change in every activity, project or plan to avoid to pass the limit global warning aforementioned.

The carbon footprint calculation tool consists of an assessment of the uses and activities to be developed in future planning that generate greenhouse gas emissions, as well as changes in land use that affect the soil's sink capacity. Mitigation or decarbonization strategies are included for assessment and quantification where data are available.

It includes the influential activities that should be included in the application for the approval of urban planning instruments, within the strategic environmental assessment procedures, in relation to the potential environmental impacts in terms of climate change.

Consequently, thanks to carbon footprint calculation tool it is possible to choose the lowest carbon emissions alternative among several and to make visible the crucial aspects that generate the most emissions at an early stage of urban development as it is a masterplan.

This guide here presented is focused on how to measure, make visible and understand the relationship between urban planning and climate change. Urbanism is planning or transforming a place, a space with identity that is the basis for human activities. The activities that need energy to be developed generate greenhouse gas (GHG) emissions. But activities are not independent of the space where they are deployed. The GHG emissions activities could not be measured independently of the place, as quantities, in an abstract way, as only space. They depend on the configuration of the space where they take place in the micro and macro scale. Therefore, depending on the location, climate, size, density, diversity or design, among others, of an environment and depending on the infrastructures, services and supplies, the GHG emissions increase or not.

This book analyses the relationship between urban planning, greenhouse gas emissions and the carbon footprint and presents the causes through the uses, activities and soil's sink capacity. Also underlines the measures to avoid emissions with quantitative and qualitative mitigation and adaptation actions and the proportion of their impact in an early stage of planning and design. Readers are provided with the knowledge needed to devise a strategy for calculating the carbon footprint of urban planning instruments for their own territories, as well as a framework for integrating climate change into the planning phase. Highlighting the importance of introduction climate change in decision-making offers a springboard for further research.

In short, the aim of this work is to make the invisible visible in order to enhance decision-making and the selection of alternatives. To have scientific data for assessment which alternative of design of an urban planning master plan (UPMP) has got a lower carbon footprint in operation, including soil's sink capacity. In other words, those alternatives that have less impact on climate change.

This tool aims to offer an instrument for measuring the carbon footprint (CF) with a tool to assess urban planning adapted as close as possible to the reality, but measurable and feasible. This is of interest so that any UPMP of a municipality or town, or a neighbourhood, that will be presented, of growth, of transformation, may be evaluated in order to adopt the better decision. There are a lot of features that are important to this, in terms of its position, its density and any other features in relation to its impact on climate change.

Currently CF calculation tools are needed to plan cities, towns and neighbourhoods in a context of climate crisis. A CF calculation tool essentially consists of an assessment of the uses and activities to be developed in future planning that generate GHG emissions, as well as changes in land use that affect the soil's sink capacity.

The carbon footprint calculation tool could help to measure different urban planning alternatives (alternative 0 or not UPMP, and others) for the urban development or transformation, as in the application of a planning proposal is shown. Consequently, thanks to carbon footprint calculation tool it is possible to choose the lowest carbon emissions alternative among several and to make visible the crucial aspects that generate the most emissions at an early stage of urban development as it is a masterplan.

The CF calculation tool that here is presented as a practical methodological design guide included all kind of uses and the derived and influential activities that should be included in the application for the approval of urban planning instruments. In the case of Community of Madrid, the region for which it has been created [1], it is used within the ordinary or simplified strategic environmental assessment procedures, in relation to the potential environmental impacts in terms of climate change, in accordance with Article 18 of Act 21/2013, of 9 December on environmental assessment. It also involves a proposal of the information to be included in the application for the initiation of the urban planning instruments subject to ordinary or simplified strategic environmental or simplified strategic environmental assessment on climate change.

The CF calculation tool has got the aim of establishing itself as a comprehensive and synthetic instrument for the quantification of GHG emissions with the ultimate aim of assessing them for decision in the current context of the climate crisis.

This carbon footprint calculation tool is applicable to the different types of urban planning, a municipality or a city or a development masterplan. This means providing practical scientific and technical advice on the assessment of climate change in urban planning.

This document shows the purpose and justification of the CF calculation tool that could be used to other territories in the case of uses or sink soil, although the data could be adjusted to the context reality. For instance, the energy mix of a region or a country could differ to other, and across time. Also, it is crucial the way the settlement is developed in a region, by the commuting could induce. Every territory shows singularities of the development model of the geographical area and climate conditions, but at the same time, in all territories cause GHG emissions the same.

It also specifies the methodological aspects of the calculation, both in its general approach with regard to energy consumption general, mobility, water consumption and waste and sewage treatment, as well as to the sink effect of soils, which varies with changes in land use as a result of planning, and finally, to GHG emissions mitigation strategies and their assessment. This tool aims to be user-friendly with very clear and available information about uses, land cover, buildability and sink capacity.

Measuring development in such important aspects as the carbon footprint helps us to make decisions in urban planning; the effects have got particular relevance because they are very long-term if not irreversible.

Furthermore, calculation fundamentals, and the carbon footprint estimation tools, case studies are presented and explained for showing the simplicity and interest to other territories.

The present carbon footprint calculation tool allows to relate planning and design with climate change through a scientific procedure.

And making the invisible visible.

Reference

1. R. Álvarez-Fernández, A. Delgado-Jiménez, F. Beltrán-Cilleruelo, Calculadora de huella de carbono para el planeamiento urbanístico de la Comunidad de Madrid. Retrieved from: https://www.comunidad.madrid/servicios/urbanismo-medio-ambiente/herramienta-hue lla-carbono-planeamiento-municipal (2022)

Contents

Abbreviations

AFV	Alternative fuel vehicle
CF	Carbon footprint (the complete accounting of greenhouse gas emissions caused directly and indirectly by human activities)
CI	Chemicals, plastics and rubber Industry
CO_{2eq}	The mass of gases emitted is measured by their CO_2 equivalence
COP	Conference of Parties
CTE-HE1	Technical Building Code, Basic Document on Energy Saving, Limitation of energy demand
CV	Coefficient of variation
DHW	Domestic hot water
EF	Emission factor
EI	Electrical, electronic and machinery Industry
FI	Food, beverages and tobacco industry
FMI	Furniture and other manufacturing industries
GHG	Greenhouse gas
GLU	Global land use
ha	Hectare
Hs	Household size
ICE	Internal combustion engine
ILU	Individual land use
$kgCO_{2eq}$	The mass of greenhouse gas emitted is measured by their CO_2 equivalence measured in kilograms
kWh	Kilowatt hour
kWh/m^3	Kilowatt hour per cubic metre
LU	Land use
m^2	Square metre
m^2c	Square metre of floor area or buildability
m^3	Cubic metre
MI	Metal industry
NI	Industry of non-metallic mineral products
PI	Printing industry and production of paper

PLD	Population of the development area
SC	Specific categories
TCI	Timber and cork industry
tCO_{2eq}/kWh	The mass of the greenhouse gas emitted is measured by their equivalence in CO_2 measured in kilogram or in ton per kilowatt hour
TI	Textile and leather Industry
UPMP	Urban planning master plan

Chapter 1
Calculation Fundamentals

There is no agreed proposal on how to build a tool that allows to evaluate the relationship between urban planning and sustainability in the city, in terms of energy supply and calculation of greenhouse gas (GHG) emissions and consequently the impact on climate change.

Carbon footprint has been historically defined as the complete accounting of GHG emissions caused directly and indirectly by human activities. Equivalent kg of carbon dioxide is the unit used to measure it, including any equivalence between CO_{2eq} and any other type of greenhouse gas for the 100-year prevalence [1].

Given the difficulty of measuring it, most quantitative studies have estimated CO_{2eq} emissions by linking them to energy consumption and quantifying emissions as a function of their carbon intensity.

The carbon footprint (CF) accounting must be based on the clear definition of the activities developed in the areas object of study and, obviously, the challenge lies in how to evaluate these greenhouse gas emissions associated with each set of activities as a result of the different emission sources implied in the performance of these activities.

This chapter resumes the theoretical approach developed as a way to join the structure of a calculation tool that takes into account the main aspects of land planning that influence in emissions and how to measure these effects. The main aspects would provide an initial diagnosis of how much carbon dioxide is emitted and a set of indicators for evaluating and controlling these emissions.

© The Author(s), under exclusive license to Springer Nature Switzerland AG 2024
R. Álvarez-Fernández et al., *A Carbon Footprint Calculation Tool for Urban Development*, SpringerBriefs in Applied Sciences and Technology,
https://doi.org/10.1007/978-3-031-69892-7_1

1.1 Framework for Carbon Footprint Calculation: Generated Emissions

The well-known report 'European Union CO_2 emissions: different accounting perspectives' summarizes the main concepts and methodologies behind three main perspectives: territorial, production and consumption-based [2]. Urban energy consumption is shaped not only by geography and economic development, but also by the history and culture of individual places. The framework of the study presented in this book focuses on how land-use planning decisions affect the emissions of the entire regional territory, caused, among other things, by activities related to the following four major specific categories (SCs):

- Transportation.
- Drinking water supply.
- Energy supply: electricity and gas supply, including renewable sources.
- Waste and sewage treatment.

The land use model obviously affects the intensity of each of the summarized activities and, consequently, the resulting GHG emissions inventory. The division of the country into homogeneous areas, defined by their geographical boundaries and their land use distribution, makes it possible to systematize and homogenize the calculation of emissions and thus to make comparisons for cases with different land use distributions.

In this sense, it is important to define the global land use (GLU) of an area as a mixture of the following individual land use (ILU) categories:

- Household.
- Tertiary (normally commercial or offices; uses which provides services to the public and includes everything from information, finance, management or insurance to entertainment).
- Facilities.
- Industrial.

Also, it is used for roads and green areas that could be private landscaped open spaces and public green areas, not only for building. In these cases, for energy supply for lighting the roads and green areas and for water supply for irrigation of green areas.

The method exposed here establishes two layers of work:

- First, a layer which includes the methodology for calculation of greenhouse gas emissions related to the different SCs.
- Secondly, a layer that identifies the group of SCs that take place in the different ILUs and finally proceed to account the carbon footprint for the global land use of a territory.

A final clarification on green areas (defined as land not designated for development). Green areas can generate GHG emissions due to maintenance and infrastructure (water consumption, public lighting, etc.), but it could be assumed that these emissions should be offset by the sequestration potential of the green areas themselves through the design of the rules that ensure the valuation of this land.

Therefore, the emission factor (EF) is defined as the average emission rate of a given GHG for a given source, relative to an activity. The carbon footprint is obtained as the sum of the multiplied terms of each emission factor (EF_i) and its expected energy consumption of the complete set of energy sources implied (C_i) in the human activity evaluated, as stated in Eq. 1.1 [3].

$$CF\left(kgCO_{2eq}\right) = \Sigma(C_i \times EF_i) \qquad (1.1)$$

When the consumption and emission factors in Eq. 1.1 are specified for each identified energy source, the mathematical relationship for determining GHG emissions expressed in Eq. 1.1 changes to a more complex function shown in Eq. 1.2.

$$CF = f\left[g_w(C_w);\; g_{ww}(G_{ww});\; g_s(G_s);\; g_p\left(C_p\right);\right.$$
$$\left. g_g\left(C_g\right);\; g_{ws}(G_{ws});\; g_t(C_t);\; EF_e, EF_g, EF_v\right] \qquad (1.2)$$

In Eq. 1.2, the value C refers to energy consumption, G is the generation value, and EF is the emission factor. The subscript w refers to water supply, ww to wastewater, e to electricity, g to gas, ws to waste and t to transportation. This general approach exposed in Eqs. 1.1 and 1.2 has been defined in previous works [4] and represents the necessary framework to cover the first methodological layer mentioned above. Each of the detailed specific activities, as well as the mathematical formulation that allows the calculation of the carbon footprint, will now be explained in more detail.

1.1.1 Water Consumption

Drinking water, potable water, or improved drinking water is a term used to describe water that is safe enough for drinking and food preparation. The water distribution infrastructure involved in the treatment and delivery of water from its initial source to its final consumption is associated with a certain amount of emissions over a given period of time (typically one year).

The carbon footprint of drinking water supply, CF_w, can be estimated by applying Eq. 1.3:

$$CF_w\left(kgCO_{2eq}/m^2\right) = EI_w\left(kWh/m^3\right) \times C_w\left(m^3/m^2\right)$$
$$\times\ EF_e\left(kgCO_{2eq}/kWh\right) \qquad (1.3)$$

where EI_w is the energy intensity of the drinking water supply, C_w represents the consumption and EF_e is the emission factor of the electricity generation mix of the country. In Eq. 1.3, the calculation of the carbon footprint has been related to a unit of area (m^2), but it could be related to human or machine consumption:

$$CF_w \left(kgCO_{2eq}\right) = EI_w \left(kWh/m^3\right) \times C_w \left(m^3\right) \times EF_e \left(kgCO_{2eq}/kWh\right) \qquad (1.4)$$

where C_w (m^3) corresponds to individual human consumption.

1.1.2 Wastewater Infrastructure

Wastewater infrastructure is the term that describes the network for the recollection, treatment and disposal of sewage and stormwater, including pipes, sewage treatment plants and outfalls among others.

The expression to link carbon footprint and wastewater management is displayed in Eq. 1.5:

$$CF_{ww} \left(kgCO_{2eq}/m_b^2\right) = EI_{ww} \left(kWh/m^3\right)$$
$$\times G_{ww} \left(m^3/m_b^2\right) \times EF_e \left(kgCO_{2eq}/kWh\right) \qquad (1.5)$$

In Eq. 1.5, the term CF_{ww} represents the wastewater carbon footprint, EF_{ww} is the emission factor referred to this concept, EI_{ww} is the energy intensity of wastewater management, G_{ww} represents the wastewater generation, and EF_e refers to the electricity generation mix emission factor.

1.1.3 Energy Supply

1.1.3.1 Electricity

The carbon footprint of the two following infrastructures, electricity and gas, can be directly rated by applying an emission factor linked to a consumption, because they are directly related to energy sources.

Thus, the carbon footprint from electricity consumption can be directly calculated by multiplying the electricity generation mix emission factor (EF_e) and the electricity consumption (C_e), as shown in Eq. 1.6.

$$CF_e \left(kgCO_{2eq}/m_e^2\right) = C_e \left(kWh/m_e^2\right) \times EF_e \left(kgCO_{2eq}/kWh\right) \qquad (1.6)$$

1.1.3.2 Gas Supply

The carbon footprint of gas consumption is obtained by applying the gas emission factor (EF_g) to gas consumption (C_g), as shown in Eq. 1.7.

$$CF_g\left(kgCO_{2eq}/m_b^2\right) = C_g\left(kWh/m_b^2\right) \times EF_e\left(kgCO_{2eq}/kWh\right) \qquad (1.7)$$

1.1.4 Waste Management Infrastructure

This term refers to the total emissions from solid waste landfills, waste incinerators, and all other waste management activities. It excludes any GHG emissions from fossil-based products (combustion or decomposition) and GHG emissions from the treatment and decomposition of organic waste.

The carbon footprint of waste management has been defined using a specific emission factor for waste management (EF_{wt}) and a waste generation rate (G_{wt}), as described in Eq. 1.8.

$$CF_{wt}\left(kgCO_{2eq}/m_b^2\right) = G_{wt}\left(t/m_b^2\right) \times EF_{wt}\left(kgCO_{2eq}/t\right) \qquad (1.8)$$

The expression to calculate the carbon footprint of waste management can also be proposed through an energy intensity factor of treating each ton of waste (EI_{wt}) as exposed in Eq. 1.9.

$$CF_{wt}\left(kgCO_{2eq}/m_b^2\right) = EI_{wt}(kWh/t) \times G_{wt}\left(t/m_b^2\right) \times EF_e\left(kgCO_{2eq}/kWh\right) \quad (1.9)$$

1.1.5 Transportation

Transportation, as a source of GHG emissions, is unavoidable in most developing countries because it plays a critical role in urban development by enabling people to access essential services: education, employment, recreation, and health care, among others. Coordination of land use and transport planning is an important measure in some countries and cities with strengthened urban institutions. Transportation planning creates great opportunities that allow citizens to achieve a more sustainable development that would benefit both the city and its inhabitants. Several problems (new or not) require new directions in transportation planning: the coming fuel shortage, climate change, health problems, space constraints, among others.

The estimation of GHG emissions from transportation is based on predictive models that are commonly used in the design of transportation infrastructure in the context of urban planning. In general, transportation results are easier to interpret than those for buildings and industry because they are not affected by regional differences in climate. It has not been necessary to rely on empirical data due to the lack of reliable information and because the predictive models commonly used in the urban planning context provide accurate and unambiguous results. Therefore, the approach to calculating the carbon footprint of transportation is presented in Eq. 1.10.

$$CF_t(kgCO_{2eq}/m_b^2) = G_t(km/m_b^2) \times EF_v(kgCO_{2eq}/km) \qquad (1.10)$$

EF_V represents a vehicle emission factor, and G_t corresponds to the covered distance during trips (due to industrial activity). EF_v must be defined based on the vehicle fleet composition and subsequent specific emission factors, as it is shown in Eq. 1.11.

$$EF_v = (r_h \times EF_h) + r_l \times \left[(r_d \times EF_d) + (r_p \times EF_p)\right] \qquad (1.11)$$

where r_h, r_l, r_d and r_p represent coefficients referred to the fleet fuel consumption in the following categories: heavy duty, light duty vehicles, gasoline consumption for on-road transportation and diesel consumption for on-road transportation rates respectively; while EF_h, EF_d and EF_p are the respective emission factors.

Finally, distance travelled Figure (G_t) in Eq. 1.9 depends on the number of trips between the analysed area i and each possible destination j, always within the corresponding influence area (N_{ij}) and along the routes of interconnection between them (d_{ij}), presented in Eq. 1.12.

$$G_t(km/m_b^2) = \sum N_{ij} \times d_{ij} \qquad (1.12)$$

Here, the transportation distances (d_{ij}) can be measured in maps, while the number of trips generated in the evaluated area and bound for all potential destinations (N_{ij}) have been estimated using a gravitational model commonly used in calculations of the carbon footprint of transport [5, 6].

Therefore, it has been assumed that the number of trips (N_{ij}) can be explained according to personal, business or leisure purposes. The expression to calculate the attraction capacity for each city is shown in Eq. 1.13.

$$N_{ij} = (r_g) \times [(r_p \times p_j/\Sigma p_j) + (r_c \times c_j/\Sigma c_j) + (r_s \times s_j/\Sigma s_j)] \times \left(1/d_{ij}^a\right) \qquad (1.13)$$

N_{ij} represents the product of the multiplication of three variables. The potential attraction of each possible destination within the influence area, which depends on the population of each possible destination as well as the number of companies and business within the analysed area. In Eq. 1.13 p_j, c_j and s_j represent the population and the number of business and shops in the jth area; r_p, r_c and r_s are the proportions of travels justified by each purpose: personal, business or leisure, respectively. There is also an impedance factor that is inversely proportional to the distance (d_{ij}) between the evaluated area and each potential destination.

There are a number of comments to be made about the methodology used to assess the carbon footprint of transport. Firstly, depending on the specific locations and transport infrastructure available, it may be appropriate to consider the alternative choice of public transport instead of the use of private vehicles. In this case, the carbon footprint of these alternatives can be calculated using the following Eqs. 1.14–1.16.

$$CF_{bus} = G_{bus}\left(km/m_b^2\right) \times EF_{bus}\left(kgCO_{2eq}/km\right) \tag{1.14}$$

$$CF_{train} = G_{train}\left(km/m_b^2\right) \times EF_{train}\left(kgCO_{2eq}/km\right) \tag{1.15}$$

$$CF_{train} = G_{train}\left(km/m_b^2\right) \times EI_{train}(kWh/km) \times EE_{train}\left(kgCO_{2eq}/kWh\right) \tag{1.16}$$

Other authors advocate for considering the alternative fuel vehicles[1] especially electric vehicles represent an alternative to conventional internal combustion engine (ICE) propulsion systems. In this case, and considering also the possibility of using public transport, Eqs. 1.9 and 1.10 should be completed with these elements as Eq. 1.17 shows.

$$CF_T\left(kgCO_{2eq}\right) = G_T \times (\%_V \times [(\%_{DS} \times EF_{DS}) + (\%_{PT} \times EF_{PT})] + (1 - \%_V)$$
$$\times [(\%_R \times IE_R \times EF_E) + ((1 - \%_R) \times EF_B)]) \tag{1.17}$$

where CF_T represents the Greenhouse gas emissions from transport, G_T the total generation of trips and $\%_V$, $\%_{DS}$, $\%_{PT}$, $\%_R$ the percentage share of trips among private vehicle, diesel, petrol and railway, respectively. Finally, EF_{DS}, EF_{PT}, EF_E and EF_B represent the emission factors of diesel, petrol, electricity generation mix and bus while IE_R refers to the energy intensity of travelling by railway.

1.2 Land Use Distribution: Initial Balance for Assessment in GHG Emissions

There are a number of basic data on the existing and planned land use situation that need to be collected in order to be able to analyse the territorial context and the type of transformation that is taking place. The goal is to know how it affects GHG emissions and especially the soil´s sink capacity.

Specifically, the basic data that need to be collected are:

---

[1] An alternative fuel vehicle (AFV) is a vehicle that uses other fuels instead of traditional petroleum fuels and also refers to any technology of powertrain that does not involve petroleum (i.e. electric car, fuel cell vehicles, solar powered).

- Figure of planning, whether it is a whole municipality or part of it, that give us a scale of the transformation to assess;
- Name of the area, if applicable, which can also be the name of the alternative to be assessed;
- Location, related to climatic zone, which will be related to the needs for heating and/or air conditioning in the different types of built uses;
- Year of calculation, which may have different emission factors according to the evolution of the same;
- The total surface area of the UPMP under assessment in carbon footprint;
- The buildability, or surface area in the plan that is assessed in climate change;
- Some population data in relation to the proposed or expected household size, either the average of the surrounding area or other to be considered. From this household size data the population of the development area is obtained. In the case of productive land uses, this will result in the floating population, which goes to the area but is not resident, and therefore affects some consumption.

Once these data have been collected, the land transformation balance of the proposal, in its initial situation, existing land use, as well as planned land use, is carried out. For each of the moments in time, the following data is collected:

- Land use, in six categories (residential, tertiary, service, commercial, industrial, undeveloped and road).
- Area of use: of each of the above, if present in the area of analysis.
- Floor area.
- Ground floor occupation, in order to know the area that remains permeable in each case.

In this way, the variation in the CO_{2eq} capture capacity of each of the soil types and, where appropriate, the vegetation cover, can be subsequently assessed. This variation (loss or gain) will be determined according to the development proposed for the area and the changes in land use that are generated.

When natural land with some vegetation cover is to be developed and converted into urban area, it will normally lose its sink capacity.

Therefore, the transformation of the existing city and brownfields instead of green-fields is recommended. And in any case, the incorporation of measures such as vegetation cover in unbuilt areas (private landscaped open spaces and public green areas) with a high CO_{2eq} absorption capacity, could improve the CF results with less GHG emissions.

1.3 GHG Mitigation and Soil's Sink Capacity: Emissions Avoided or Captured

The CF calculation tool has got different possibilities to show the measures avoid or capture GHG emissions.

After the analysis of what generates GHG emissions (different uses located in buildings and energy consumption in roads and green areas), it is key that any tool for measuring the impact of climate change in urban planning could make visible what is working as a sinkhole. Then, in this case it is not the use or activity that generates the emissions itself, but rather the change in land use from the initial balance to the proposed design for urban planning. This land cover change affects the absorption capacity of CO_{2eq} or sink, which implies a different vegetation cover of the land where the urban planning is implemented.

This means considering in the carbon footprint the change in land use itself where the UPMP is implanted, which varies its capacity as a sink, and not only the emissions due to uses and activities to be developed.

The changes in land use from the initial situation to the UPMP scenario are therefore considered in order to assess how the carbon sink functions in this aspect.

The CO_{2eq} capture of the vegetation cover of the area is analysed, differentiating between tree species: poplar, conifer, conifer and other species, irrigated herbaceous crops, dry crops, shrubs, shrubs with conifer, shrub with another non-coniferous species and so on up to a total of 28 possible covers with the differences in CO_{2eq} capture depending on the research taken as a source. In the final situation, public green areas and private landscaped open spaces are also considered as permeable surface of the area.

The urbanized surface is considered non-permeable and in the event that strategies are followed to make it permeable and therefore there is no soil sealing, i.e. sustainable urban drainage systems, it can be considered a mitigation measure to be assessed as other measures, although it is not quantified.

The issue is to take into account permeable soil, CO_2 fixation in the initial and final state. It is known that soil is a scarce and non-renewable resource. Preserving permeable soil, especially with vegetation cover, improves the CO_2 sink function. In this way it is possible to prioritise whether an alternative is more rigorous and maintains areas with trees with a certain function as a sink, compared to another that does not take this into account.

In addition to quantifying the activities that generate emissions, as seen in the previous point, it is also possible to quantify the emissions that have been avoided. These are directly quantified in the cases of percentage of energy consumption using renewables, percentage of reclaimed water or separate waste treatment.

In the case of the alternatives shown as case studies in Chap. 3, self-sufficiency has been assessed, which implies significant variations on business-as-usual case study.

It is also important to include measures that are not quantifiable but which it was important to evaluate positively in a carbon footprint report that will accompany urban planning in its environmental processing. To achieve this goal, there are free fields for introducing mitigation and adaptation measures beyond those that can be quantitatively assessed today. Also for new measures that may emerge in the future.

Such examples can be the number of recharging points for battery electric vehicles, the inclusion of a heat network, walkability, among others.

In this last example, it is in this very positive and currently so often quoted concept of the 15-min city, which proposes that most of citizens need services (such as work, shopping, education, health care or leisure) that could be located within walking or cycling distances of less than 15 min from any point in the city [7]. This strategy design foster soft mobility and avoid emissions of private vehicles. In the event that these mitigation and adaptation measures can be measured, these values are introduced and reduce the carbon footprint. In any case, even if the reduction is not known, it is important to declare it for comparing alternatives.

The action and the effort that is being made, for example, to choose between two alternatives, is positively valued.

Finally, it is important to underline that at this stage of the urban planning the uncertainty is total, as only the distribution of land according to its use has been carried out with its key factors and the mitigation and captures measures proposed, as adaptation, in its case. Therefore, the tool serves to make approximations that will have the usefulness of allowing a comparison between two or more urban development solutions.

Crucially, a carbon footprint-based climate change assessment cannot evaluate a transformation carried out on an area of high vegetation cover in the same way as on land that does not have this CO_2 absorption capacity. Because if we only analyse activities, this crucial issue of the territory would not be taken into account.

This is an important point for the CF calculation tool, not only to review activities but soil's sink capacity variation due to urban transformations, because its importance in the long term.

References

1. IPCC, *Climate Change 2007: Synthesis Report. Contribution of Working Groups I, II and III to the Fourth Assessment Report of the Intergovernmental Panel on Climate Change* (Intergovernmental Panel on Climate Change (IPCC), Geneva, 2007).
2. EEA, EEA Technical Report No 20/2013 European Union CO_2 emissions: different accounting perspectives (2013)
3. H.S. Eggleston, L. Buendia, K. Miwa, T. Ngara, K. Tanabe (eds.), *2006 IPCC Guidelines for National Greenhouse Gas Inventories* (IPCC National Greenhouse Gas Inventories Programme, Hayama, 2006)
4. S. Zubelzu, A. Hernández, Methodolgy for household carbon footprint calculation incorporated in urban planning procedures, in *Proceedings of XVIII International Congress on Project Management and Engineering*, 16–17 July 2014
5. S. Zubelzu, R. Álvarez, Urban planning and industry in Spain: A novel methodology for calculating industrial carbon footprints. Energy Policy **83**, 57–68 (2015)
6. S. Zubelzu, R. Álvarez, A, Hernández Methodology to calculate the carbon footprint of household land use in the urban planning stage. Land Use Policy **48**, 223–235 (2015)
7. C. Moreno, The 15-Minute City: redesigning urban life with proximity to services. Barcelona Societatl *Journal on Social Knowledge and Analysis*, (2024)

Chapter 2
Carbon Footprint Estimation Tool

The tool for calculating greenhouse gas emissions generated by urban planning is developed in an Excel workbook consisting of 11 main tabs (Fig. 2.1):

- Content and instructions: index of the calculation tool, together with the colour criteria for filling in the cells.
- Basic data: contains the basic and general data of the current planned development.
- Household: data related to plots of land for household use are entered.
- Commercial (Tertiary): data related to the plots, whose use is commercial, offices, etc., are entered.
- Public facilities: data related to plots whose use is public are entered.
- Industrial: data related to plots for industrial use are entered.

  And also it includes:

- Road_green_areas: data related to public lighting and irrigation of public green areas are entered.
- Sink_mitigation: data related to land use conditions (sink) and other mitigation measures are entered.
- Result: results are collected by source and use.
- Report: summary of the new planning data for printing.
- Remarks: explanatory notes for correct completion.

2.1 Basic Data

This is the first tab of the tool where the user will start entering basic information that the tool will use to calculate the greenhouse gas emissions generated by urban planning.

As explained in this chapter, this tab is divided into two sections. In the first one, basic data related to the study area and its population are entered, such as:

© The Author(s), under exclusive license to Springer Nature Switzerland AG 2024
R. Álvarez-Fernández et al., *A Carbon Footprint Calculation Tool for Urban Development*, SpringerBriefs in Applied Sciences and Technology,
https://doi.org/10.1007/978-3-031-69892-7_2

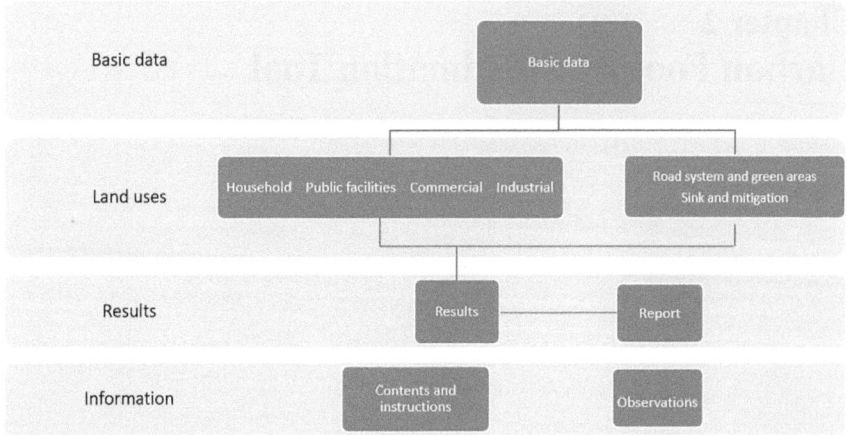

Fig. 2.1 Structure of the tool for calculating greenhouse gas emissions generated by urban planning
Note: Commercial means all kind of tertiary uses

- Figure of planning: differentiating between masterplan or development plan.
- Generic data of the area: name of the area, municipality, year, zone climate (this could be auto-completed, depending on the municipality selected), surface area (auto-completed, it represents the sum of the area of use of each of the types of use in the planned land, data to be filled in the second part of this sheet) and built-up surface (autocompleted, represents the sum of the area of the different types of use on the planned land, data to be filled in the second part of this sheet).
- Population study: the tool takes as a value for occupants per dwelling of 2.5 inhabitants and can be modified if the user has a real value of the study area.

The population of the development area refers to the estimated population in residential use. This cell is auto-completed once the data has been entered in the household tab, as shown in Eq. 2.1.

$$PLD_T(\text{inhabitants}) = Hs(\text{inhabitants/household}) * \sum_{i=1}^{n} \text{dwellings}_i(ud) \quad (2.1)$$

where PLD represents the population of the development area (related to residential use), Hs is the household size and dwellings represent the number of dwellings in each of the sectors of the area for residential use.

There exist a floating population associated with public facilities, commercial and industrial uses. This cell is auto-completed once the data has been entered in the public facilities, commercial and industrial tab, as shown in Eq. 2.2.

$$\begin{aligned}
FP_T(\text{inhabitants}) = {}& \left[d_{PF}(\text{inhabitants}/m_c^2) \times A_{PF}(m_c^2)\right] \\
& + \left[d_C(\text{inhabitants}/m_c^2) \times A_C(m_c^2)\right] \\
& + \left[d_I(\text{inhabitants}/m_c^2) \times A_I(m_c^2)\right]
\end{aligned} \tag{2.2}$$

where FP_T is the population associated with public facilities, commercial and industrial uses, d_{PF} represents the density of floating population in public facilities, A_{PF} is the buildable area of plots for public facilities, d_C represents the density of floating population in tertiary uses, A_C is the buildable area of plots for tertiary uses, d_I represents the density of floating population in industrial uses, A_I is the buildable area of plots for industrial uses (Fig. 2.2; Table 2.1).

In the second part of the Table, a description of the area is presented, differentiating between the current status and the planned proposal. Surface area data will be entered for each type of use, differentiating among six different uses that the tool allows the land to be used for: household, commercial, public facilities, industrial, undeveloped and road.

For each one of these, the total area, buildable area and ground floor area will be entered, as proposed in Fig. 2.3.

2.2 Land Use

Although the tool is based on a specific methodology for the calculation of greenhouse gas emissions described in this point, the tool allows the user to enter the values of CO_{2eq} emissions if they have been calculated using a different method to the one explained in this book for each of the four land sources studied.

Figure 2.4 shows the case of knowing the emissions generated by waste management in commercial use. The way to enter these values is the same for each use and source of generation.

2.2.1 Household

2.2.1.1 Electricity and Gas Supply

Energy consumption can be calculated according to the particularities of each land use. A distinction can be made between a first level of influence that corresponds to the consumption of energy for air conditioning and domestic hot water and the

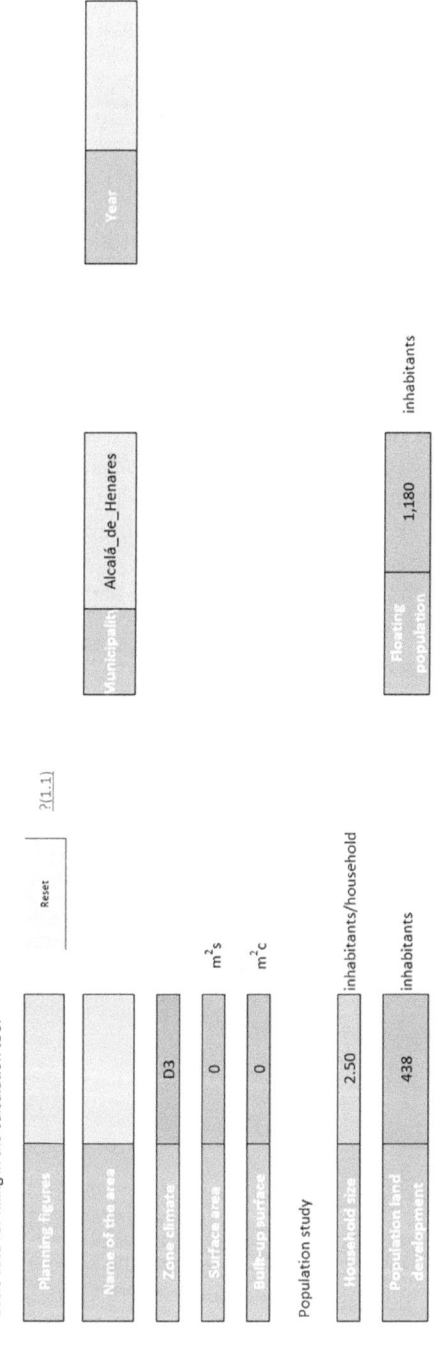

Fig. 2.2 First part basic data tab design

Table 2.1 Floating population density[1]

Tertiary (Commercial, offices)	0.04	Inhabitants/m^2c
Public facilities	0.02	Inhabitants/m^2c
Industrial	0.02	Inhabitants/m^2c

second level of influence that represents the energy consumption for the performance of other activities.

In the first level, energy consumption can be established by means of the energy certification label of the projected or existing building, taking into account the particularities of each land use, established by public or private institutions, as it is in Spain for example the Institute for the Diversification and Saving of Energy (IDAE). In turn, it is considered that the electrical energy network serves household appliances, lighting and air conditioning, while heating and hot water are supplied by natural gas. On the other hand, the calculation can also be carried out by applying the Technical Building Code, approved in 2006, allowing the percentages of gas and electricity to be distributed in different ways.

On the second level, energy consumption must be estimated for the performance of the activities carried out in each land use, which will be different in each one, both in terms of the quantities and the type of fuel used. The values for the energy consumption of this second level in the residential, industrial, commercial and service sectors are obtained from various sources available in the literature [1, 2].

Air Conditioning and Domestic Hot Water (DHW)

The tool also allows the calculation of CO_{2eq} emissions based on electricity demand according to the typology and climate zone in order to comply with the Technical Building Code (CTE-HE1) or based on electricity demand according to the energy certification.

Option 1. According to Typology and Climate Zone

The thermal demand per unit area (TD_i) is the energy used for air conditioning and DHW for each dwelling. Depending on the use of the land under study, this will vary.

$$TD_i\left(kWh/m_c^2\right) = E_{ACS}\left(kWh/m_c^2\right) + E_{CAL}\left(kWh/m_c^2\right) + E_{REF}\left(kWh/m_c^2\right) \quad (2.3)$$

---

[1] The objective is to determine the number of trips per built-up area. In the case of the Community of Madrid and under normal conditions this information is based on empirical studies; frequently the generation rates used in traffic are: Commercial (tertiary): 0.04 trips/ m^2 built (15% heavy vehicles); Industrial: 0.014 trips/m^2 built (80% heavy vehicles); Installations: 0.016 trips/ m2 built area.

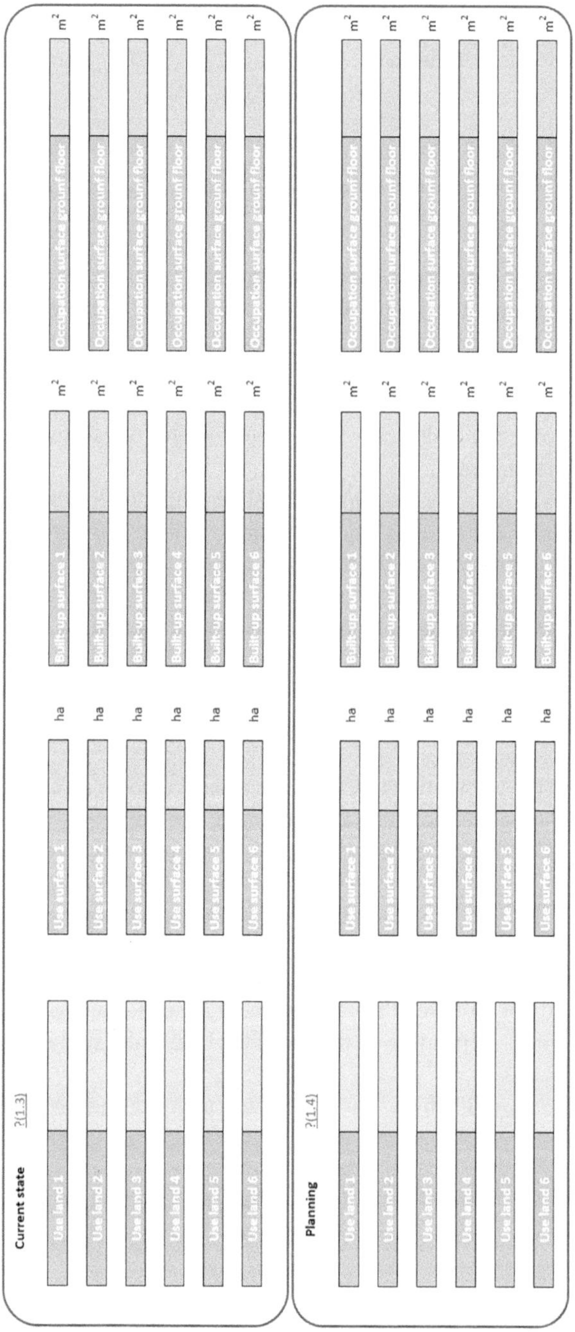

Fig. 2.3 Second part basic data tab design

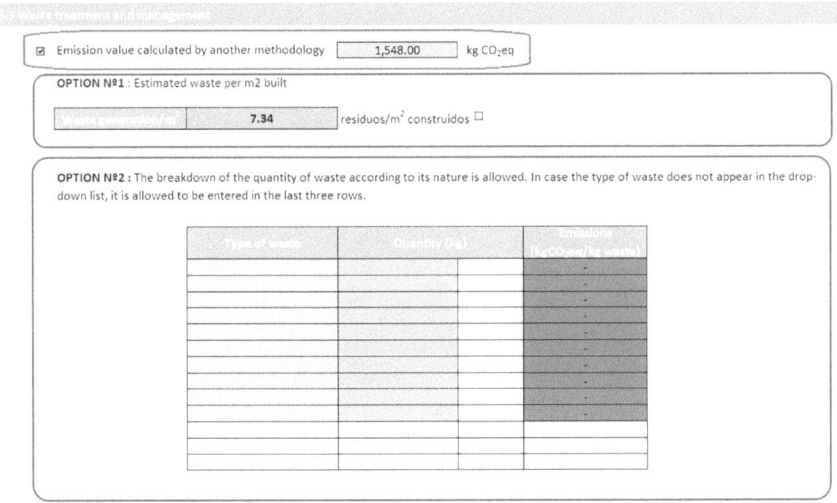

Fig. 2.4 Example of entering emissions calculated by other methodology

where E_{ACS} is the energy required for DHW per unit area, E_{CAL} represents the energy required for heating per unit area and E_{REF} is the energy required for cooling per unit area.

This is a variable that depends on the type of housing (single-family or multi-family) and the climate zone where the urban development under study is analysed. The tool determines the climate zone between E1, D1 and E3, depending on the altitude of the municipality according to the CTE.[2]

$$TD = f(\text{climate zone, type of housing}) \qquad (2.4)$$

The total calculation of emissions as a function of thermal demand is expressed by Eq. 2.5.

$$CF_{AC\text{-}T}(kgCO_2eq) = \sum TD_i(kWh/m_c^2) \times A_i(m_c^2) \times \%EM(kgCO_2eq/kWh) \qquad (2.5)$$

where $CF_{AC\text{-}T}$ represents the CO_{2eq} emissions on all plots, TD_i is the thermal demand of each of the plots, A_i is the total buildable area of each of the plots and %EM is the percentage of energy mix (between the energy sources electricity and gas) (Fig. 2.5).

The tool allows the study of two types of buildings within the residential land use such as single-family and multi-family dwellings.

---

[2] Certificado de Eficiencia Energética [3].

Option nº1 ☑

Electrical energy demand according to typology and climate zone to comply with the Technical Building Code (CTE-HE1).

Percentage GAS	100%		Percentage ELECTRICAL	0%

	Area	Type of dwellings	Thermal demand (kWh/m²c)		Buildability (m²c)		Buildability/house (m²c/dwelling)	Number of dwellings		Total landscaped open spaces of the area	
2(2.2)	Area 1	Multi-family	44.00	kWh/m²c	10,000.00	m²c	200 m²c/dwelling	50	dwellings	1,000.00	m²
	Area 2	Multi-family	44.00	kWh/m²c	15,000.00	m²c	120 m²c/dwelling	125	dwellings	1,000.00	m²
			-		-		-	0	-	-	
			-		-		-	0	-	-	
			-		-		-	0	-	-	
					25,000.00			175		2,000.00	

Example:

	Area	Type of dwellings	Thermal demand (kWh/m²c)		Buildability (m²c)		Buildability/house (m²c/dwelling)	Number of dwellings		Total landscaped open spaces of the area	
	Area 1	Single-family	53.00	kWh/m²c	10,000.00	m²c	85 m²c/dwelling	118	dwellings	1,000.00	m²
	Area 2	Multi-family	53.00	kWh/m²c	15,000.00	m²c	m²c/dwelling	150	dwellings	1,000.00	m²

Fig. 2.5 Example of data entry in the tool. Section: air-conditioning and DHW (compliance with CTE-HE1) option 1 household use

Table 2.2 Electrical and thermal demand according to climate zone and building type. Household use[4]

Building type	Climate zone	Electrical demand (kWh/m²c)	Thermal demand (kWh/m²c)	Total (kWh/m²c)
Single-family	D1	33	53	86
Single-family	D3	33	54	87
Single-family	E1	33	63	96
Multi-family	D1	33	43	76
Multi-family	D3	33	44	77
Multi-family	E1	33	53	86

Table 2.3 Energy mix by energy source[5]

Energy source	Energy mix (kgCO₂eq/kWh)
Gas	0.198
Electricity	0.241

Furthermore, given the diversity of dwellings that can be found in the same area, differentiated by their quality or their built surface, among other variables, the tool allows to differentiate up to six sub-areas within an area, each one with its own characteristics. These entered data will be used for the estimation of emissions for each of the sources analysed by this tool.

An average buildability per dwelling of 100 m² is estimated,[3] but this can be varied (Table 2.2).

On the other hand, the tool makes it possible to determine the percentage of the energy source used for air-conditioning and DHW in dwellings, differentiating between gas and electricity (Table 2.3).

Option 2. According to Their Energy Certification

The tool also allows the determination of CO_{2eq} emissions from the value of the label of the energy certification, which is used for the building planning each of the plots under study.

$$CF_{AC-T}(kgCO_2eq) = \sum GE_i(kgCO_2eq/m_c^2) \times A_i(m_c^2) \qquad (2.6)$$

---

[3] In urban planning in Spain, one dwelling per 100 m² of residential use is usually calculated for urban standards and the development of urbanization projects. As indicated in the text, this figure can be varied depending on the type of urban transformation that needs to be assessed.

[4] Gobierno de España [4].
   Accessed 15/03/2024.

[5] IEA International Energy Agency, https://www.iea.org/countries/spain.
   Accessed 17/03/2024.

Table 2.4 Emissions according to energy rating

Energy certificate	GE_i	
A	1.75	$kgCO_{2eq}/m^2$
B	5.00	$kgCO_{2eq}/m^2$
C	8.80	$kgCO_{2eq}/m^2$
D	14.40	$kgCO_{2eq}/m^2$
E	27.95	$kgCO_{2eq}/m^2$
F	40.70	$kgCO_{2eq}/m^2$
G	43.20	$kgCO_{2eq}/m^2$

CF_{AC-T} represents the CO_{2eq} emissions of each of the plots, GE_i[6] represents the global emission of each of the plots according to their energy certificate and A_i is the total buildable area of each of the plots (Table 2.4).

Energy Consumption. Other Uses of Electricity

For the calculation of GHG emissions due to energy consumption according to other uses (DE), the consumption associated with cooling/heating and DHW of inhabited places is disregarded, as this has been taken into account in the previous point (compliance with CTE-HE1 or according to energy certification).

For the calculation of CO_{2eq} emissions, it is possible to make an estimate of the average energy consumption according to the use of the activity within each type of use (Option 1) or to calculate the emissions according to a declaration of energy sources within a list of fuel types (Option 2).

$$DE\left(kWh/m_c^2\right) = E_{ILUM}\left(kWh/m_c^2\right) + E_{EQ}\left(kWh/m_c^2\right) \tag{2.7}$$

where E_{ILUM} is the energy required for lighting per unit area and E_{EQ} is the energy required for different equipment per unit of area.

Option 1. According to Estimated Consumption

$$CF_{E-T}(kgCO_2eq) = \sum EC_i(kWh/dwellings) \times dwellings_i(ud)$$
$$\times EM(kgCO_2eq/kWh) \tag{2.8}$$

where CF_{E-T} represents the total CO_{2eq} emissions on the plots, EC_i is the estimated electricity consumption of each of the dwellings, $dwellings_i$ is the estimated number of dwellings and ME is the energy mix of the country.

---

[6] Certificados energéticos [5]
  Accessed 15/02/2024.

The number of dwellings is calculated by the tool as explained in Sect. 2.1.1. Air conditioning and DHW (compliance with CTE-HE1 or according to energy certification) of the tool, either estimating a buildability per dwelling of 100 m^2 or allowing the user to modify the buildability per dwelling based on A_i which is the total built area of each of the plots (Fig. 2.6).

$$\text{Dwellings}_T = \sum A_i / (\text{Buildability}/\text{dwelling})(\text{m}^2_{\text{dwelling}}/\text{dwelling}) \qquad (2.9)$$

where dwelling s_T represents the total number of dwellings, A_i is the buildability of each of the plots and buildability/dwelling is the estimated buildability per dwelling.

The estimation of electricity consumption is calculated according to compliance with CTE-HE1 or according to the energy certification. The choice of one or the other is made in the previous point, 2.1.1 Air conditioning and DHW (compliance with CTE-HE1 or according to energy certification) (Fig. 2.7).

The tool allows to determine how much of the energy consumed comes from self-consumption in both options. This mitigation of emissions from self-consumption will be reflected in the Mitigation Section, which will be defined later.

Option 2. According to the Declaration of the Energy Source

At this point and by means of a statement of the current consumption of different energy sources, the total CO_{2eq} emissions are obtained.

$$CF_{E\text{-}T}\left(\text{kgCO}_2\text{eq}\right) = \sum CR_i \times EF_i(\text{kgCO}_2\text{eq}/ud) \qquad (2.10)$$

where $CF_{E\text{-}i}$ represents the CO_{2eq} emissions in each of the plots, CR_i is the actual consumption of each different fuel type and EF is the fuel emission factor (Fig. 2.8; Table 2.5).

2.2.1.2 Drinking Water Supply

The average annual water use on different types of land can be calculated from various sources of information [7–9] either in terms of square metres built or per inhabitant. Once these consumptions have been determined, their relationship to emissions can be analysed through the energy cost associated with the water cycle. This covers the processes of adduction, water purification, distribution, sewerage, purification and regeneration. This cost is related to the energy required expressed in kWh per cubic metre of water supplied [7] and together with the emission factor of the electricity mix in Spain averaged over the last five years [6] allows to calculate the carbon footprint of water consumption.

CO_{2eq} emissions from drinking water consumption on residential plots are divided into emissions from multi-family and single-family dwellings, as a result of the difference in consumption between the two types of dwellings.

Option nº2 ☑
Electricity demand according to your energy certification

Area	Energy certification	Global emissions (kgCO₂-eq/m²c)		Buildability (m²c)		Buildability/house (m²c/dwelling)		Number of dwellings		Total landscaped open spaces of the area	
2(2.3)											
Area 1	B	5.00	kgCO₂/m²c	10,000.00	m²c	125	m²c/dwelling	80	dwellings	1,000.00	m²
Area 2	C	8.80	kgCO₂/m²c	15,000.00	m²c	125	m²c/dwelling	120	dwellings	1,000.00	m²
		-		-		-		0	-	-	-
		-		-		-		0	-	-	-
		-		-		-		0	-	-	-
		-		-		-		0		-	-
				25,000.00				200		2,000.00	

Example:

Area 1	D	14.40	kgCO₂/m²c	10,000.00	m²c	125	m²c/dwelling	80	dwellings	500.00	m²
Area 2	C	8.80	kgCO₂/m²c	10,000.00	m²c		m²c/dwelling	100	dwellings	800.00	m²

Fig. 2.6 Example of data entry in the tool. Section: air-conditioning and DHW (compliance with CTE-HE1) option 2 household use

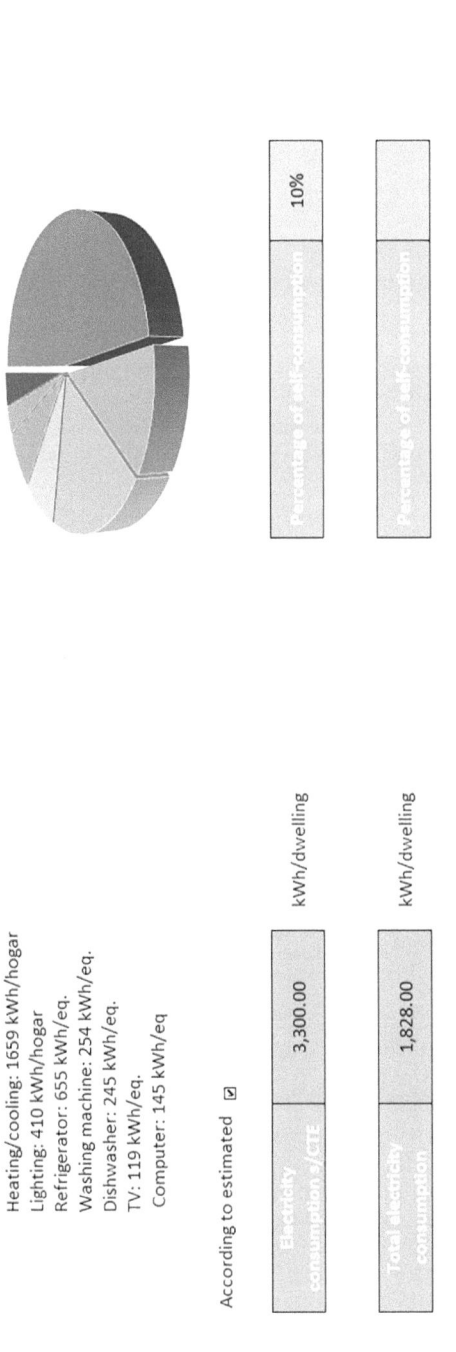

Fig. 2.7 Example of data entry in the tool. Section 2.1.2 Energy consumption other uses Electricity, option 1 Household use

Change in emissions by energy source declaration

?(3.5)	Fuel types	Emissions (kgCO₂eq)		Total quantity	
	Natural gas	2.15	kg CO₂eq/Nm³ natural gas	100.00	Nm³ natural gas
	Propane gas	2.94	kg CO₂eq/kg propane gas	100.00	kg propane gas
		0.00	-		-
		0.00			
		0.00	-		-
		0.00	-		-

Fig. 2.8 Example of data entry in the tool. Section 2.1.2 Energy consumption other electricity uses, option 2. All uses

Table 2.5 Emission factors according to fuel type[7]

Fuel type		Emission factors	
Electricity	kWh	0.241	kg CO_{2eq}/kWh
Natural gas	(m^3)	2.15	kg CO_{2eq}/Nm^3 natural gas
Butane gas	kg	2.96	kg CO_{2eq}/kg butane gas
Butane gas (cylinders)	number	37.06	kg CO_{2eq}/ butane gas (cylinders)
Propane gas	kg	2.94	kg CO_{2eq}/kg propane gas
Propane gas (cylinders)	number	102.84	kg CO_{2eq}/ propane gas (cylinders)
Diesel oil	litres	2.79	kg CO_{2eq}/litre diesel oil
Fuel oil	kg	3.05	kg CO_{2eq}/kg fuel oil
Generic LPG	kg	2.96	kg CO_{2eq}/kg generic LPG
Domestic coal	kg	2.3	kg CO_{2eq}/kg domestic coal
Imported coal	kg	2.53	kg CO_{2eq}/kg imported coal
Petroleum coke	kg	3.19	kg CO_{2eq}/kg petroleum coke

$$CF_W(kgCO_2eq) = [CF_w]_{\text{single-family}}(kgCO_2eq) + [CF_w]_{\text{multi-family}}(kgCO_2eq)$$

(2.11)

where CF_w represents the total CO_{2eq} emissions of residential plots associated with water consumption, $CF_{w\ \text{single-family}}$ is the total estimate of emissions associated with water consumption in single-family dwellings and $CF_{w\ \text{multi-family}}$ is the total estimate of emissions associated with water consumption in multi-family dwellings.

The energy per water cycle in the Community of Madrid is (0.57 kWh/m^3).[8] The cycle comprises the processes of adduction, water purification, distribution,

---

[7] Oficina Catalana de Cambio Climático [6]

[8] Fundación Canal [9].
   Accessed 12/03/2024.

sewerage, purification and regeneration. Source: Energy footprint in the integral water cycle in the Community of Madrid [9].

It is estimated that the average annual supply is 198 L/inhabitant/day, corresponding to domestic consumption 126 L/inhabitant/day, i.e. 64%, and to community consumption 72 L/inhabitant/day, the remaining 36%.

$$CF_{W \text{ single-family}}(kgCO_2eq) = \left[CW_s\left(m^3\right) \times E_{CW}\left(kWh/m^3\right) \times EM\ (kgCO_2eq/kWh) \right]_{\text{single-family}}$$
(2.12)

$$CW_s\left(m^3\right) = \left[\text{Capacity}\ (ud) \times \text{Ratio}\left(m^3/\text{people}\right)\right]_{\text{single-family}}$$
(2.13)

where CF_w represents the CO_{2eq} emissions in single-family dwellings associated with water consumption, CW_s is the estimated total water consumption in single-family dwellings (198 L/inhabitant/day), E_{CW} is the energy per water cycle of the Community of Madrid and EM is the energy mix.

For the calculation of water consumption (CW_u) the average household size is estimated to be (2.5 inhabitants/dwelling)[9] (Basic data tab of the tool), which together with the average consumption and the number of single-family dwellings gives an estimate of total consumption (see Eq. 2.9).

$$CF_{W \text{ multi-family}}(kgCO_2eq) = \left[CW_m\left(m^3\right) \times E_{CW}\left(kWh/m^3\right) \times EM\ (kgCO_2eq/kWh) \right]_{\text{multi-family}}$$
(2.14)

$$CW_m\left(m^3\right) = \left[\text{Capacity}\ (ud) \times \text{Ratio}\left(m^3/\text{people}\right)\right]_{\text{multi}-\text{family}}$$
(2.15)

where CF_w represents the CO_{2eq} emissions in multi-family dwellings associated with water consumption, CW_m is the estimated total water consumption in multi-family dwellings (126 L/inhabitant/day) per number of multi-family dwellings, E_{CW} is the energy per water cycle of the Community of Madrid and EM is the energy mix (Fig. 2.9).

2.2.1.3 Waste and Sewage Treatment

This section includes direct and indirect emissions during the entire management process: collection and transport, transfer plants, pre-treatment plants, final treatment plants and final disposal of waste. In addition, a distinction is made between the possibility of separate collection or not (see Fig. 2.10).

The different types of waste generated by each land use and the corresponding greenhouse gas emissions can be calculated from data obtained from various sources [9, 11, 12].

---

[9] Comunidad de Madrid [10].
Accessed 15/03/2024.

For the calculation of water demand, a distinction is made between single-family dwellings with a garden and all other dwellings.

| Household consumption | 126.00 | litres/inhabitant per day |

| Community consumption (single- | 72.00 | litres/inhabitant per day |

| Energy water cycle | 0.57 | kWh/m^3 |

| Number of single-family houses with garden | 50 | dwellings | ?(2.5)

| Other dwellings | 125 | dwellings |

Fig. 2.9 Water consumption. Household use

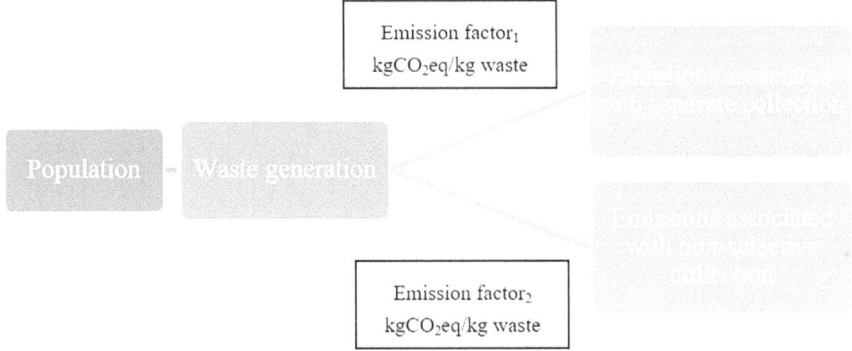

Fig. 2.10 Methodology for calculating emissions in the waste treatment and waste management process

For the estimation of CO_{2eq} as a consequence of the generation of municipal solid waste, the tool differentiates between emissions generated in a selective collection or not.

As a starting point, it is necessary to estimate the number of inhabitants in the area. This information has been calculated in the previous point, and the estimated mass (kg) of waste generated by each individual.

The following equation (Eq. 2.16) is then applied.

$$CF_{WTG}(kgCO_2eq) = PAD(\text{inhabitants}) \times CTG_{unitary}(\,kg_{residue}/\text{inhabitants})$$
$$\times FE\,(kgCO_2eq/kg\ residue) \qquad (2.16)$$

In this equation CF_{WTG} represents the CO_{2eq} emissions associated with waste generation and treatment, PAD the population of the development area under study, $CTG_{unitary}$ is the estimate of waste generated per inhabitant and FE is the greenhouse gas emission factor.

The information needed to fill in the parameters in the equation must be sought in studies published by official bodies, such as, in example, a city council or a region. The estimate of waste generated per inhabitant in this case was obtained from official data collected in the study of Generation and composition of household and commercial waste in the Community of Madrid by fractions (2016) (Table 2.6).

As the tool differentiates between areas with separated collection and areas without separated collection, two types of emission factors must be differentiated:

- $EF_1\,(kgCO_{2\,eq}/kg\ residue)$ (greenhouse gas emissions factor in separated collection)[10]

---

[10] Guía técnica para la medición, estimación y cálculo de emisiones al aire. Valorización de residuos sólidos urbanos. Retrieved from:

https://www.euskadi.eus/contenidos/documentacion/eprtr/es_guia/adjuntos/residuos_urbanos.pdf

Table 2.6 Distribution of the estimated waste generation per inhabitant[11]

Waste	Type of waste	Gross kg/inhabitant	Net kg/inhabitant	Gross (%)	Net (%)
Light packaging	Packaging	39.47	28.82	10.06	7.35
Organic matter	Form	73.72	73.72	18.79	18.79
Garden and pruning waste	Form	53.65	53.65	13.68	13.68
Cellulose	Paper	22.33	16.94	5.69	4.32
Textiles	Rest	30.65	25.35	7.81	6.46
Non-packaging wood	Rest	10.73	10.73	2.74	2.74
Wood Commercial/Industrial Packaging	Rest	2.7	2.7	0.69	0.69
Glass (packaging)	Glass	22.64	22.64	5.77	5.77
Non-packaging plastics (except refuse bag film)	Rest	8.03	7.52	2.05	1.92
Waste bag film	Rest	7.31	3.01	1.86	0.77
Plastic Commercial/Industrial Packaging (except C/I film)	Packaging	1.27	1.2	0.32	0.31
Film Commercial/Industrial	Rest	1.67	1.42	0.43	0.36
Minor construction debris	Rest	13.05	13.05	3.33	3.33
Steel non-packaging	Rest	1.82	1.82	0.46	0.46
Steel Commercial/Industrial packaging	Packaging	0.1	0.1	0.03	0.03
Aluminium non-packaging	Rest	0.1	0.1	0.03	0.03
Aluminium packaging Commercial/Industrial	Packaging	0	0	0.00	0.00
Multimaterials	Rest	10.19	10.191	2.60	2.60
WEEE	Rest	4.31	4.31	1.10	1.10
Unclassifiable material	Rest	24.43	24.43	6.23	6.23
Other	Rest	5.75	5.75	1.47	1.47
Paper/cardboard	Paper	58.32	52.2	14.87	13.31
TOTAL	Packaging	392.24	359.651	100,00	91.69

- EF_2 ($kgCO_{2eq}$/kg residue) (greenhouse gas emissions factor without separated collection.)[12]

Accessed 15/03/2024.

[11] Consejería de Medio Ambiente, Administración Local y Ordenación del Territorio. Comunidad de Madrid [13]. Estudio de generación y composición de residuos domésticos en la Comunidad de Madrid 2016. Retrieved from: https://www.comunidad.madrid/sites/default/files/doc/medio-amb iente/presentacion_del_estudio_de_residuos_domesticos_en_la_cm.pdf
Accessed 10/02/2024.

[12] Guía práctica para el cálculo de emisiones de gases de efecto invernadero (GEI) OFICINA CATALANA DE CAMBIO CLIMATICO. Retrieved from:

2.2.1.4 Transportation

The rapid growth of greenhouse gas emissions from public and private transport is inevitable in most developing countries, as transport plays a vital role in development.

Transport planning creates enormous opportunities for a more sustainable society, benefiting cities and their residents. There are several issues that require new directions in transport planning: upcoming fuel shortages, climate change, health concerns, space constraints and among others.

The estimation of total greenhouse gas emissions from transport is based on effective prediction models commonly used in transport infrastructure design in the context of urban planning. Therefore, the method for calculating the carbon footprint of transport is summarized in Eq. 2.17.

$$CF_{M\text{-total}}(kgCO_2eq) = G_{t\text{-}i}(km) \times EF_{v\text{-}i}(kgCO_2eq/km) \tag{2.17}$$

where $CF_{M\text{-total}}$ represents the total CO_{2eq} emissions associated with mobility, $G_{t\text{-}i}$ is the total distance travelled during trips (by all vehicle types) and $EF_{v\text{-}i}$ is the emission factor of the vehicles, which will depend on the fuel type.

Depending on the type of transport, the emission factor will be different. There are three main categories: private transport, public transport and freight transport by heavy duty vehicles. Each of them interferes in a different way depending on the use of the study area, but it is possible to establish a framework based in the following Eqs. (2.18–2.24).

$$CF_{M\text{-private diesel}}(kgCO_2eq) = \left[Gt_i(km) \times EF_{v\text{-diesel}}(kgCO_2eq/km)\right] \tag{2.18}$$

$$CF_{M\text{-private diesel}}(kgCO_2eq) = \left[Gt_i(km) \times EF_{v\text{-petrol}}(kgCO_2eq/km)\right] \tag{2.19}$$

$$CF_{M\text{-public train}}(kgCO_2eq) = \left[Gt_i(km) \times EF_{v\text{-train}}(kgCO_2eq/km)\right] \tag{2.20}$$

$$CF_{M\text{-public bus}}(kgCO_2eq) = \left[Gt_i(km) \times EF_{v\text{-bus}}\left(\frac{kgCO_2eq}{km}\right)\right] \tag{2.21}$$

$$CF_{M\text{-public underground}}(kgCO_2eq) = \left[Gt_i(km) \times EF_{v\text{-underground}}(kgCO_2eq/km)\right] \tag{2.22}$$

$$CF_{M\text{-heavy vehicle lorry}}(kgCO_2eq) = \left[Gt_i(km) \times EF_{v\text{-heavy vehicle lorry}}(kgCO_2eq/km)\right] \tag{2.23}$$

https://descubrelaenergia.fundaciondescubre.es/files/2013/07/Guia-practica-calcul-emisiones_rev_ES.pdf
Accessed 15/03/2024.

$$CF_{\text{M-heavy vehicle van}}\left(\text{kgCO}_2\text{eq}\right) = \left[Gt_i(\text{km}) \times EF_{\text{v-heavy vehicle van}}\left(\text{kgCO}_2\text{eq/km}\right)\right]$$
$$(2.24)$$

where Gt_i represents the distance travelled by each type of transport mode from each of the source populations to the final population. As it can be expected, this is the most difficult variable to determine.

To determine the mobility generated by the sector or area under planning, it is necessary to know the data on the number of trips (journeys) caused as a result of the activity in each land use planned in the study area and to analyse where they occur. This data could be available in reports at the web sites of agencies and through Official Publications of the European Communities.

In the case presented in this book, the data comes mainly from the Comunidad de Madrid government and other public or private research sources.

Based on these data, the theory of Gravitational Modelling has been applied, which enables the distribution of trips from a population (origin) to a set of populations (destinations), analysing the various possible options depending on the level of attraction (Fig. 2.11). It is a factor that is subject to the mass (inhabitants) of the various destination towns and the corresponding distances from the origin, which in turn symbolizes the distances from the origin. Based on this information, the number of trips to each of the destinations, and hence the total kilometres travelled, can be calculated.

Figure 2.11 shows the equation of Newton's Law of Universal Gravitation, which has been used as the basis for the development of the gravitational models. In these models, it is considered that a population with a larger number of inhabitants exerts a greater attraction for reasons of forced mobility. On the other hand, the greater distance between towns has a deterrent effect for various reasons (travel time, fuel costs, etc.).

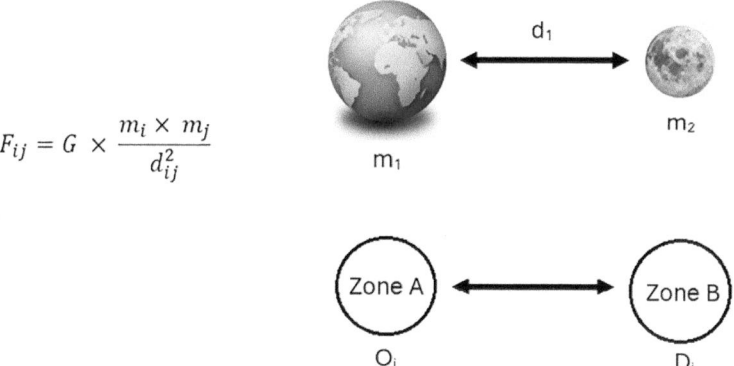

$$F_{ij} = G \times \frac{m_i \times m_j}{d_{ij}^2}$$

Fig. 2.11 Newton's Law of Universal Gravitation as a basis for the development of gravitational models

What is the way to calculate the probability that a person living in A will move to each of the municipalities? To do this, the authors have created what they define as attraction matrices, which are the basis for the application of the Universal Gravitational Law to mobility within the region studied (in this case the Community of Madrid).

These matrices are derived from the implementation of the Universal Gravitational Law in each of the 179 municipalities of the Community of Madrid, with the purpose of determining the probability that a person travels from one municipality to another. A probability that is determined as a function of the population of each of the municipalities and their distance.

The total distance travelled can be calculated by knowing how many trips each person makes and where they are most likely to go.

The connection between kilometres and emissions is established through the distribution of trips and workforces. The statistical analysis of the percentage composition of the vehicle fleet (diesel and petrol) in the Community of Madrid is attached using the Dirección General de Tráfico and the estimated fossil fuel consumption of the different vehicle models.

Emissions from public transport journeys are also tested using the emission factors of these means of transport per kilometre and passenger [6]. First, an estimated value is established for the number of heavy goods vehicles and vans in the industrial and commercial sectors, as well as the sum of kilometres travelled on average by lorries and vans. The values have been established on the basis of various studies [6].

The developers of the tool have established that not all areas produce the same type of mobility. Therefore, two types of attraction matrices have been calculated. The first matrix (output matrix) relates to residential areas, where the population tends to move to their jobs or leisure, and a second matrix (input matrix), linked to the endowment, commercial and industrial areas, because it is suggested that these areas attract mobility.

Calculation of the Mobility Attraction Matrix 'Out'

Each of the elements of this matrix will determine the probability of a person moving from a municipality of origin to a destination municipality (Fig. 2.12).

The Community of Madrid agglutinates made up of 179 municipalities. Therefore, the square matrix is of order 179. For the application of the Universal Gravitational Law, the population of each of these municipalities and the distance between them must be taken as a starting point (see where the data comes from).

Taking as a starting point the equation of the Universal Gravitational Law:

$$F_{ij} = \left[G \times \left(M_i \times M_j \right) \right] / r_{ij}^2 \qquad (2.25)$$

Fig. 2.12 Matrix of attraction 'out' Community of Madrid

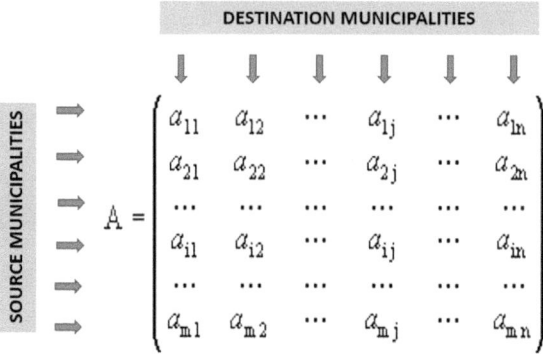

where F_{ij} is the force of attraction between two bodies, G is the universal gravitational constant, M_1 is the mass of body 1, M_2 is the mass of body 2 y r_{ij} is the distance between the two bodies.

Considering that each row represents the relationship between a source municipality and the remaining 179 municipalities, Eq. 2.25 can be adapted and simplified to a mobility model.

$$a_{ij} = P_{\text{destination}}(\text{inhabitants})/r^2_{\text{source destination}}(\text{km}) \tag{2.26}$$

where a_{ij} is the force of attraction between two municipalities, $P_{\text{destination}}$ is the population of the destination municipality and $r_{\text{source-destination}}$ is the distance between the municipality of origin and destination.

To analyse the attraction that occurs within the same municipality, the distance $r_{\text{source-destination}}$ is calculated as the equivalent surface of the municipality to that which would occupy a circumference of the same area and the radius of the same is taken as the calculation distance.

Applying Eq. 2.26 to determine the probability of mobility between two municipalities with respect to each other, Eq. 2.27 is applied to each element of the matrix to determine the probability of mobility between two municipalities with respect to each other.

$$a_{ij}(\%) = \left(a_{ij}/\sum_{j=1}^{179} a_{ij} \right) \tag{2.27}$$

The type of mobility that occurs in residential areas is an 'out' mobility, i.e. the population of the area tends to move to their jobs or leisure activities. Therefore, the data for attraction between municipalities calculated in the 'out' attraction matrix will be used.

According to the work (Comunidad de Madrid 2018), the number of trips per inhabitant is (0.46 trips/inhabitant * day). Therefore, if the probability of where

Table 2.7 Distribution of trips according to the size of the municipality[13]

| | Ministerio de Fomento | | | | |
	Population	Private vehicle (%)	Public transport (%)	Walking (%)	Bicycle
Minor	$\leq 100,000$	47.00	13.60	37.40	2.00
Medium	$10,000 < X < = 500,000$	43.00	12.20	43.10	1.70
Large	$> 500,000$	24.60	42.60	31.30	1.50

these trips can go and the number of them is known, the total number of km travelled can be obtained.

$$Gt_{total}(km) = \sum_{j=1}^{179}(a_{ij} \times P_{development} \times 0.46_{trips/inhab} \times r_{i-j}) \qquad (2.28)$$

where Gt_{total} is the total distance travelled by the population of the area under study, a_{ij} is the force of attraction between two municipalities, $P_{development}$ is the population of the area under study and r_{i-j} is the distance between the municipality of origin and the destination.

It is true that Gt_{total} includes journeys with all possible means of transport, each with different emission factors. According to the data collected, the distribution of these journeys according to the type of means of transport used is related to the size of the population (Table 2.7).

$$Gt_{total-private}(km) = Gt_{total} \times \% \text{ use} \qquad (2.29)$$

$$Gt_{total-public}(km) = Gt_{total} \times \% \text{ use} \qquad (2.30)$$

where $Gt_{total-private}$ is the total distance travelled by the population of the study area by private means of transport, $Gt_{total-public}$ is the total distance travelled by the population of the study area by public means of transport, % use is the percentage of each means of transport (as shown in Tables 2.7 and 2.8.

By applying Eqs. 2.18, 2.19, 2.20, 2.21 and 2.22 with their respective emission factors to the distances travelled according to the means of transport, the amount of CO_{2eq} emissions is estimated in the mobility section (Table 2.9).

The tool allows varying the percentages allocated to public or private transport as well as the distribution of vehicle distribution according to their engine capacity/emissions (Fig. 2.13).

---

[13] IDAE. Vehicle information section.
https://coches.idae.es/base-datos/intervalo-de-emisiones

Table 2.8 Distribution of vehicles according to their engine capacity/emissions[14]

		Distribution of vehicles according to their engine capacity/emissions		
		diesel (emission factors)		
	Type A	76.80%	150	gCO₂eq/km
	Type B	13.70%	166	gCO₂eq/km
	Type C	9.50%	121	gCO₂eq/km
		Gasoline (emission factors)		
	Type D	79.00%	145	gCO₂eq/km
	Type E	11.00%	111	gCO₂eq/km
	Type F	10.00%	212.8	gCO₂eq/km

Table 2.9 Emission factor

	Emission factor (g CO₂eq/ passenger-km)
Underground	30
Train	33
Bus	49

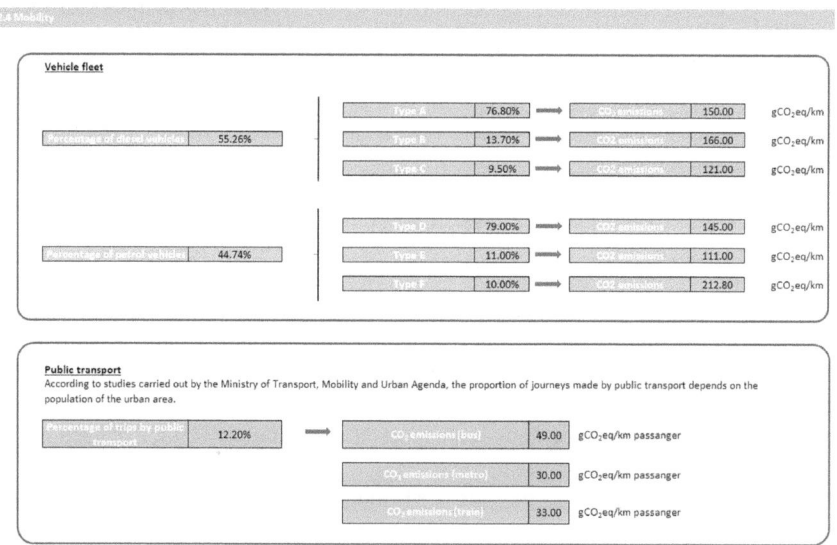

Fig. 2.13 Example mobility data household use

---

[14] IDAE. Vehicle information section.
   https://coches.idae.es/base-datos/intervalo-de-emisiones

Table 2.10 Electricity and heat demand according to climate zone and type of use: Tertiary use[15]

Type of use	Climate zone	Electrical demand (kWh/m²c)	Thermal demand (kWh/m²c)	Total (kWh/m²c)
Administrative	D1	92	97	189
Administrative	D3	92	158	250
Administrative	E1	92	115	207
Recreational	D1	251	129	380
Recreational	D3	251	169	420
Recreational	E1	251	148	399
Commercial	D1	79	68	147
Commercial	D3	79	99	178
Commercial	E1	79	83	162
Hotel	D1	466.2	569.8	1036
Hotel	D3	466.2	569.8	1036
Hotel	E1	466.2	569.8	1036

2.2.2 Tertiary: Commercial and Offices

As a first step in estimating the emissions generated in a commercial use environment, the sub-areas for commercial use must be defined and differentiated by their sub-category.

Given the diversity of commercial subcategories (as shown in Table 2.10) that can occur within an urban sector and the different demands of each of them, the tool offers the possibility to enter up to 10 different sub-areas in each sector.

For each area the following variables must be entered: name, subcategory, plot area (m²), buildability coefficient, landscaped open spaces (m²), the variables buildability (m²c) and occupation are calculated directly by the tool (Fig. 2.14).

These entered data will be used for the estimation of emission calculations for each of the sources analysed by this tool.

2.2.2.1 Electricity and Gas Supply

For the calculation of CO_{2eq} emissions based on the demand for electricity for air conditioning and DHW, the methodology is based on Eqs. 2.5 and 2.6, explained in the previous points (2.2.1 Household).

---

[15] IDAE. Energy balances. https://www.idae.es/en/information-and-notifications/studies-reports-and-statistics/statistics-and-energy-balance

Accessed 20/03/2024.

Description of plots for tertiary use in planning according to their subcategory and surface area

Area	Subcategory tertiary	Plot area (m²)		Buildability coefficient		Buildability (m²c)		Landscaped open spaces (m²)		Occupation
Area 1	Administrative	10,000.00	m²	0.80	m²/m²	8,000.00	m²c	1,000.00	m²	320
Area 2	Commercial	5,000.00	m²	0.85	m²/m²	4,250.00	m²c	500.00	m²	170
		-	-	-	-	-	-	-	-	
		-	-	-	-	-	-	-	-	
		-	-	-	-	-	-	-	-	
		-	-	-	-	-	-	-	-	
		-	-	-	-	-	-	-	-	
		15,000.00	m²		-	12,250.00	m²c	1,500.00	m²	490

Fig. 2.14 Description of plots for commercial use in planning according to their subcategory and surface area

In studies of urban developments for commercial use, a distinction can be made between four different types of activities: administrative, recreational, commercial and hotel.

As in the case of residential use, the tool calculates the emissions in two ways: Option 1 (according to typology and climate zone) and Option 2 (according to the declaration of the energy source) (Fig. 2.15).

Energy Consumption Other Uses Electricity

Option 1. According to Estimated Consumption

It is the result of a sum of consumptions. See Eq. 2.23.

$$CF_{E\text{-}T}\left(kgCO_2eq\right) = \sum ED_i(kWh/m^2c) \times A_i\left(m^2\right) \times EM\ (kgCO_2eq/kWh)$$
(2.31)

Option nº1 ☑
Electrical energy demand according to typology and climate zone to comply with the Technical Building Code (CTE-HE1).

?(3.3)	Area	Subcategory tertiary	Thermal demand (kWh/m²c)		Buildability (m²c)	
	Area 1	Administrative	158.00	kWh/m²c	8,000.00	m²c
	Area 2	Commercial	99.00	kWh/m²c	4,250.00	m²c
				-		-
				-		-
				-		-
				-		-
				-		-
				-		-

Fig. 2.15 Example of data entry in the tool. Section: Air-conditioning and DHW (compliance with CTE-HE1) option 1. Commercial use

Option nº 2 ☑
Electricity demand according to your energy certification

?(3.4)	Area	Energy certification	Global emissions (kgCO₂eq/m²c)		Buildability (m²c)	
	Area 1	B	5.00	kgCO₂/m²c	8,000.00	m²c
	Area 2	C	8.80	kgCO₂/m²c	4,250.00	m²c
				-		-
				-		-
				-		-
				-		-
				-		-
				-		-

Fig. 2.16 Example of data entry in the tool. Section: Air conditioning and DHW (compliance with CTE-HE1) option 2. Commercial use

where CF_{E-T} represents the CO_{2eq} emissions on each of the plots, ED_i is the electricity demand of each of the plots (as shown in Table 2.10, column electrical demand (kWh/m^2_c)), A_i is the total area of each of the plots and EM is the energy mix.

Option 2. According to the Declaration of the Energy Source

For the calculation of CO_{2eq} emissions based on the demand for electricity for other uses, the methodology is based on Eq. 2.10, explained in the previous points (2.2.1 Household).

2.2.2.2 Drinking Water Supply

Total emissions are divided between emissions resulting from water consumption for the activity of the use and those associated with water consumption for irrigation.

$$CF_{W \text{ commercial-use}}\left(kgCO_2eq\right) = [\ CF_w]_{\text{activity}}\left(kgCO_2eq\right) + [\ CF_w]_{\text{irrigation}}\left(kgCO_2eq\right) \tag{2.32}$$

where $CF_{w \text{ tertiary-use}}$ represents the CO_{2eq} emissions on commercial plots associated with water consumption, $CF_{w \text{ acivity}}$ is the total estimate of emissions associated with the activity and $CF_{w \text{ irrigation}}$ is the total estimate of emissions associated with irrigation.

$$CF_{W \text{ activity}}\left(kgCO_2eq\right) = CW_{\text{activity}}\left(m^3\right) \times E_{CW}\left(kWh/m^3\right) \times EM\ \left(kgCO_2eq/kWh\right) \tag{2.33}$$

where $CF_{W \text{ activity}}$ represents the CO_{2eq} emissions associated with the water consumption associated with the activity, CW_{activity} is the total estimated water consumption associated with the activity, E_{CW} is the energy per water cycle of the Community of Madrid and EM is the energy mix.

For the estimation of $CW_{\text{activity}}\left(m^3\right)$ (0.32 m^3/m^2_c) is taken as a reference value; in the event that a real value of water consumption of the activity is available, this can be modified.

To determine the water consumption associated with irrigation, the tool allows an estimate to be made if the type of irrigation of the free landscaped areas is unknown, or it allows the type of irrigation used on the area to be defined: (Fig. 2.17; Table 2.11).

Default calculation if irrigation type is unknown. The tool takes the average turf watering value:

$$CF_{W \text{ irrigation}}\left(kgCO_2eq\right) = S_{\text{green-areas}}\left(m^2\right) \times Q\left(m^3/m^2year\right)$$
$$\times E_{CW}\left(kWh/m^3\right) \times ME\ \left(kgCO_2eq/kWh\right) \tag{2.34}$$

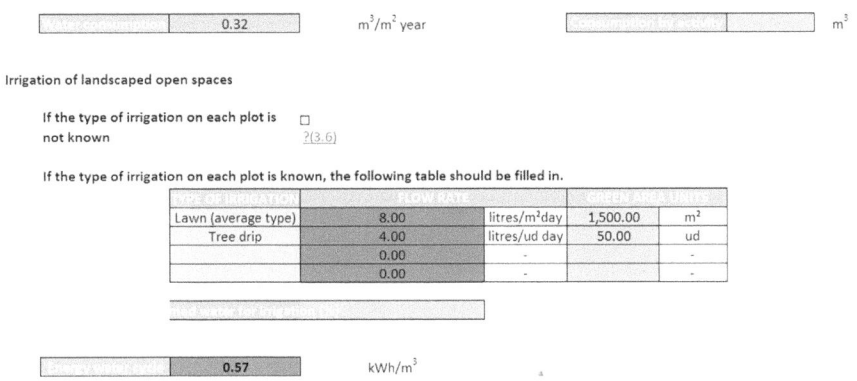

Fig. 2.17 Example of data entry in the tool. Section: Water consumption, Irrigation of landscaped open spaces. Tertiary and facilities

Table 2.11 Average flow rates according to type of irrigation[16]

Type of irrigation	Average flow rate	
Lawn (average type)	8	litres/m²day
Shrub drip	8	8.00 L/hour (15 min per day)
Tree drip	16	16.00 L/hour (15 min per day)

where $S_{green\text{-}areas}$ is the free landscaped area (starting point for the Description of plots), Q is the water flow for average lawn irrigation, E_{CW} is the energy per water cycle of the Community of Madrid and ME is the energy mix.

In case the type of irrigation is known:

$$CF_{W\ irrigation}(kgCO_2eq) = \sum S_{i\text{-green-areas}}(m^2) \times Q_i(m^3/m^2a\,n\,o)$$
$$\times E_{CW}(kWh/m^3) \times EM(kgCO_2eq/kWh) \qquad (2.35)$$

where $S_{i\text{-green areas}}$ is the free landscaped area of each type, Q_i is the water flow required in each type of irrigation, E_{CW} is the energy per water cycle of the Community of Madrid and EM is the energy mix.

2.2.2.3 Waste and Sewage Treatment

Given the wide variety of activities that can take place in commercial or service uses, the tool, in addition to estimating emissions based on the square metres built, allows

---

[16] Statistical Yearbook Community of Madrid.
   https://www.madrid.org/iestadis/fijas/estructu/general/anuario/ianuserie.htm

OPTION Nº2 : The breakdown of the quantity of waste according to its nature is allowed. In case the type of waste does not appear in the drop-down list, it is allowed to be entered in the last three rows.

Type of waste	Quantity (kg)		Emissions (kgCO₂eq/kg waste)
Light packaging	1,000.00	kg	0.12
Glass (packaging)	1,500.00	kg	0.031
Organic matter	2,500.00	kg	0.393
			-
			-
			-
			-
			-

Fig. 2.18 According to waste generation declaration

the quantity of waste generated of each type to be entered if a declaration of actual waste generation is available.

Based on an Estimate Per Square Metre Built

$$CF_{WTG}(kgCO_2eq) = A(m^2) \times CTG_{unitary}(kg_{residuo}/m_c^2)$$
$$\times EF \ (kgCO_2eq/kg \ residue) \tag{2.36}$$

where CF_{WTG} represents the CO_{2eq} emissions associated with waste generation and treatment, A is the square metres built, $CTG_{unitary}$ is the estimate of waste generated per square metre and EF is the greenhouse gas emission factor.

The waste generation per square metre is estimated to be $(7.34 \ kg_{residuo}/m_c^2)$[17]

According to Waste Generation Declaration

$$CF_{WTG-T}(kgCO_2eq) = \sum CRSU_i \times EF_i(kgCO_2eq/kg \ de \ residue) \tag{2.37}$$

where CF_{WTG-T} represents the CO_{2eq} emissions associated with waste generation and treatment, $CRSU_i$ is the actual amount of each type of waste and EF_i is the emission factor depending on the type of waste (Fig. 2.18; Table 2.12).

---

[17] GUÍA PRÁCTICA PARA EL CÁLCULO DE EMISIONES DE GASES DE EFECTO INVER-NADERO (GEI). Retrieved from: https://descubrelaenergia.fundaciondescubre.es/files/2013/07/Guia-practica-calcul-emisiones_rev_ES.pdf
    Accessed 10/04/2024.

Table 2.12 Emission factors according to waste type[18]

Waste	Emissions (kg CO_{2eq}/kg waste)
Light packaging	0.12
Organic matter	0.393
Garden and pruning waste	0.393
Cellulose	0.056
Textiles	0.625
Non-packaging wood	0.625
Wood Commercial/Industrial Packaging	0.625
Glass (packaging)	0.031
Non-packaging plastics (except refuse bag film)	0.625
Waste bag film	0.625
Plastic Commercial/Industrial Packaging (except C/I film)	0.12
Film Commercial/Industrial	0.625
Minor construction debris	0.625
Steel non-packaging	0.625
Steel commercial/Industrial packaging	0.12
Aluminium non-packaging	0.625
Aluminium packaging Commercial/Industrial	0.12
Multimaterials	0.625
WEEE	0.625
Unclassifiable material	0.625
Other	0.625
Paper/cardboard	0.056

2.2.2.4 Transportation

Gravity models can also be used for commercial use. The plots for commercial use produce movements of attraction from other municipalities, in contrast to those originating in plots for residential use, which produce movements to other municipalities. Therefore, in order to estimate the emissions generated by mobility, an 'in' mobility matrix is created.

---

[18] Gobierno Vasco.
https://www.euskadi.eus/contenidos/documentacion/eprtr/es_guia/adjuntos/combustion.pdf
Accessed 15/03/2024.

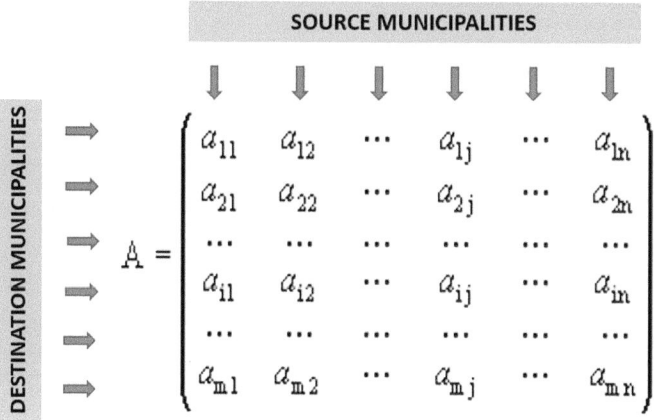

Fig. 2.19 Matrix of attraction 'in' Community of Madrid

Calculation of the Mobility Attraction Matrix 'in'

The methodology used for the calculation of the 'in' mobility attraction matrix is very similar to that used for the 'out' mobility attraction matrix. This matrix represents with which attraction an area attracts mobility from other municipalities (Fig. 2.19).

Considering that each row represents the relationship between a destination municipality and the remaining 179 municipalities, Eq. 2.25 can be adapted and simplified to a mobility model.

$$a_{ij} = P_{\text{source}}(\text{inhabitants})/r^2_{\text{source-destination}}(\text{km}) \qquad (2.38)$$

where a_{ij} is the force of attraction between two municipalities, $P_{\text{destination}}$ is the population of the destination municipality and $r_{\text{source-destination}}$ is the distance between the municipality of origin and destination.

To analyse the attraction that occurs within the same municipality, the distance $r_{\text{source-destination}}$ is calculated as the equivalent surface of the municipality to that which would occupy a circumference of the same area and the radius of the circumference is taken as the calculation distance.

By applying Eq. 2.38, a value is obtained for each municipality of origin to each municipality of destination, to determine the probability of mobility between two municipalities with respect to the rest, Eq. 2.39 is applied to each element of the matrix.

$$a_{ij}(\%) = \left(a_{ij} \middle/ \sum_{j=1}^{179} a_{ij} \right) \qquad (2.39)$$

Table 2.13 Percentage of trips in heavy and light vehicles by domains[19]		Light vehicles (%)	Heavy vehicles (%)
	Commercial use	85.00	15.00

Table 2.14 Estimated trips in the commercial sectors[20]		Estimated trips
	Commercial use	0.04 trips/m²built

In the study of greenhouse gas emissions caused by mobility in commercial areas, an additional variable must be introduced, namely emissions from heavy-duty vehicles (trucks and vans) (Table 2.13).

As in the case of a built-up area, commercial areas can be considered as centres of attraction for mobility, hence the 'in' attraction matrix should be used for the estimation. On this occasion, the estimation of trips differs from the previous areas.

The estimate of total kilometres travelled by private light-duty vehicles and public transport is obtained from the following expression.

$$Gt_{\text{total-light}}(\text{km}) = \sum_{j=1}^{179}\left(a_{ij} \times E_{\text{use}}\left(\text{viajes/m}^2_{\text{construido}}\right) \times A_{\text{built}} \times \%\text{vehiculo ligero} \times r_{i-j}\right)$$

(2.40)

where $Gt_{\text{total-light}}$ is the total distance travelled by the population of the area under study in private vehicles and public transport, a_{ij} is the force of attraction between two municipalities (matrix 'in'), E_{use} is the estimate of trips per built − up area (as shown in Table 2.14), A_{built} is the built-up area of the plot, % light vehicle is the percentage of light vehicles and y r_{i-j} is the distance between the municipality of origin and destination. The distribution of vehicles according to their engine capacity/emissions, as well as the percentage of public and private transport, is determined in the same way as in the household and public facilities area (as show in Fig. 2.13).

In order to determine the kilometres travelled by heavy vehicles and the different emission factors according to their typology, the authors of the tool use the following ratios determined in studies of the Community of Madrid for their estimation.

$$Gt_{\text{total-heavy}}(\text{km}) = \left(\text{Ratio}_1\left(\text{van/m}^2_{\text{built}}\right) \times A_{\text{built}} \times \text{annual distance}_1\right)$$
$$+ \left(\text{Ratio}_2\left(\text{trucks/m}^2_{\text{built}}\right) \times A_{\text{built}} \times \text{annual distance}_2\right) \quad (2.41)$$

where $Gt_{\text{total-heavy}}$, is the total distance travelled by heavy vehicles, Ratio_1, is the ratio of vans/day per 1000 m²built for the Community of Madrid (as shown in Table 2.15),

---

[19] Statistical Yearbook Community of Madrid.
   https://www.madrid.org/iestadis/fijas/estructu/general/anuario/ianuserie.htm

[20] Statistical Yearbook Community of Madrid.
   https://www.madrid.org/iestadis/fijas/estructu/general/anuario/ianuserie.htm

Table 2.15 Ratio of heavy goods vehicles according to the use of the area and the type of heavy goods vehicle

	Commercial use
Ratio of van/day per 1000 m^{2}c	2.50
Ratio of trucks/day per 1000 m^{2}c	1.00

Table 2.16 Estimated mileage of a heavy goods vehicle in the Community of Madrid[21]

Heavy vehicle. Van	42,000.00 km/year
Heavy vehicle. Trucks	57,500.00 km/year

Table 2.17 Emission factors according to heavy duty vehicle type

		Emission factors (g CO_{2eq} /km)
Van	Petrol	220.36
Van	Diesel	282.47
Diesel truck	Rigid < 14 t	490.73
Diesel truck	Rigid > 14t	663.01
Diesel truck	Articulated < 34t	579.96
Diesel truck	Articulated > 34t	791.44

Ratio$_2$ is the ratio of vans/day per 1000 m$^2_{built}$ for the Community of Madrid (as shown in Table 2.15), annual distance$_1$ is the annual distance travelled by a van for the Community of Madrid (as shown in Table 2.16), annual distance$_2$, is the annual distance travelled by a truck for the Community of Madrid (as shown in Table 2.16), and A_{built} is the built-up area of the plot (Table 2.17).

Applying Eqs. 2.23 and 2.24 with their respective emission factors to the distances travelled depending on the means of transport, the amount of CO_{2eq} emissions in the mobility section is estimated.

The attached image represents the heavy goods vehicles section within the mobility section in an industrial use for the Community of Madrid (Fig. 2.20).

2.2.3 Facilities

As a first step in estimating the emissions generated in an area of public facilities use, the sub-areas for public facilities use must be defined and differentiated by their sub-category (as shown in Table 2.18).

Given the diversity of sub-areas that can be found within an urban development sector and the different demands of each of them, the tool offers the possibility of introducing up to 10 different sub-areas in each sector.

---

[21] Statistical Yearbook Community of Madrid.
    https://www.madrid.org/iestadis/fijas/estructu/general/anuario/ianuserie.htm

Heavy vehicles   ?(3.7)
The weight of each type of heavy goods vehicle must be estimated for the activity in question.

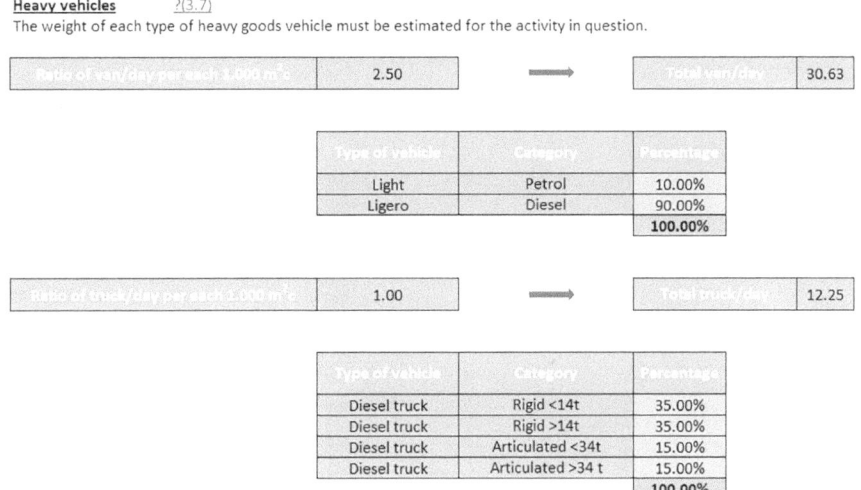

| Ratio of van/day per each 1.000 m²c | 2.50 | ⟶ | Total van/day | 30.63 |

Type of vehicle	Category	Percentage
Light	Petrol	10.00%
Ligero	Diesel	90.00%
		100.00%

| Ratio of truck/day per each 1.000 m²c | 1.00 | ⟶ | Total truck/day | 12.25 |

Type of vehicle	Category	Percentage
Diesel truck	Rigid <14t	35.00%
Diesel truck	Rigid >14t	35.00%
Diesel truck	Articulated <34t	15.00%
Diesel truck	Articulated >34 t	15.00%
		100.00%

Fig. 2.20 Example of the distribution of heavy goods vehicles in the Community of Madrid in the commercial use

For each area the following variables must be entered: name, subcategory public facilities, plot area (m²), buildability coefficient, landscaped open spaces (m²), the variables buildability (m²$_c$) and occupation are calculated directly by the tool.

These entered data will be used for the estimation of emission calculations for each of the sources analysed by this tool (Fig. 2.21)

2.2.3.1 Electricity and Gas Supply

For the calculation of CO_{2eq} emissions based on the demand for electricity for air conditioning and DHW, the methodology is based on Eqs. 2.5 and 2.6, explained in the previous points (2.2.1 Household).

In the study of urban developments destined for public use, a distinction can be made between six different types of activities: public administration, sports, educational, cultural, hospital and residential.

As in the case of residential use, the tool calculates the emissions in two ways: Option 1 (according to typology and climate zone) and option 2 (according to the declaration of the energy source) (Figs. 2.22 and 2.23).

Table 2.18 Electricity and heat demand according to climate zone and type of use as facilities[22]

Type of use	Climate zone	Electrical demand (kWh/m²c)	Thermal demand (kWh/m²c)	Total (kWh/m²c)
Public administration	D1	92	97	189
Public administration	D3	92	158	250
Public administration	E1	92	115	207
Sport	D1	27	76	103
Sport	D3	27	76	103
Sport	E1	27	82	109
Educational	D1	23	84	107
Educational	D3	23	84	107
Educational	E1	23	107	130
Cultural	D1	92	97	189
Cultural	D3	92	158	250
Cultural	E1	92	115	207
Hospital	D1	60	244	304
Hospital	D3	60	287	347
Hospital	E1	60	293	353
Retirement homes	D1	58	171	229
Retirement homes	D3	58	178	236
Retirement homes	E1	58	207	265

Energy Consumption Other Uses Electricity

Option 1. According to Estimated Consumption

$$\mathrm{CF_{E\text{-}T}}\left(\mathrm{kgCO_2eq}\right) = \sum \mathrm{ED}_i(\mathrm{kWh/m^2c}) \times A_i\left(\mathrm{m^2}\right) \times \mathrm{EM}(\mathrm{kgCO_2eq/kWh})$$

$$(2.42)$$

where $\mathrm{CF_{E\text{-}T}}$ represents the $\mathrm{CO_{2eq}}$ emissions on each of the plots, ED_i is the electricity demand of each of the plots (as shown in Table 2.18, column electrical demand (kWh/m2c)), A_i is the total built area of each of the plots and EM is the energy mix.

---

[22] Statistical Yearbook Community of Madrid.
   https://www.madrid.org/iestadis/fijas/estructu/general/anuario/ianuserie.htm

Description of plots for dotacional use in planning according to their subcategory and surface area

Area	Subcategory dotacional	Plot area (m²)		Buildability coefficient		Buildability (m²o)		Landscaped open spaces (m²)		Occupation
Area 1	Sport	5,000.00	m²	0.80	m²/m²	4,000.00	m²c	800.00	m²	80
Area 2	Educational	4,000.00	m²	0.80	m²/m²	3,200.00	m²c	600.00	m²	64
Area 3	Cultural	3,500.00	m²	0.80	m²/m²	2,800.00	m²c	500.00	m²	56
			-		-		-		-	
			-		-		-		-	
			-		-		-		-	
			-		-		-		-	
			-		-		-		-	
		12,500.00	m²			10,000.00	m²c	1,900.00	m²	200

Fig. 2.21 Description of plots for public facilities use in planning according to their subcategory and surface area

Option nº1 ☑
Electrical energy demand according to typology and climate zone to comply with the Technical Building Code (CTE-HE1).

?(4.3)

Area	Subcategory tertiary	Thermal demand (kWh/m²c)		Buildability (m²c)	
Area 1	Sport	76.00	kWh/m²c	4,000.00	m²c
Area 2	Educational	84.00	kWh/m²c	3,200.00	m²c
Area 3	Cultural	158.00	kWh/m²c	2,800.00	m²c
			-		-
			-		-
			-		-
			-		-
			-		-
			-		-
			-		-

Fig. 2.22 Example of data entry in the tool. Section: Air-conditioning and DHW (compliance with CTE-HE1) Option 1. Facilities

Option nº 2 ☑
Electricity demand according to your energy certification

?(4.4)

Area	Energy certification	Global emissions (kgCO₂eq/m²c)		Buildability (m²c)	
Area 1	A	1.75	kgCO₂/m²c	4,000.00	m²c
Area 2	B	5.00	kgCO₂/m²c	3,200.00	m²c
Area 3	B	5.00	kgCO₂/m²c	2,800.00	m²c
			-		-
			-		-
			-		-
			-		-
			-		-
			-		-

Example:

Area a	D	14.40	$kgCO_2/m^2c$	15,000.00	m^2c

Fig. 2.23 Example of data entry in the tool. Section: Air conditioning and DHW (compliance with CTE-HE1) Option 2. Facilities

Option 2. According to the Declaration of the Energy Source

For the calculation of CO_{2eq} emissions based on the demand for electricity for other uses, the methodology is based on Eq. 2.10, explained in the previous point, see (2.2.1 Household).

2.2.3.2 Drinking Water Supply

For the calculation of CO_{2eq} emissions based on water consumption, the methodology is based on Eqs. 2.32–2.35, explained in the previous points. See (2.2.2 Tertiary: commercial and offices).

Table 2.19 Electrical and thermal demand depending on climate zone and type of use. Industrial use

Type of use	Climate zone	Electrical demand (kWh/m²c)	Thermal demand (kWh/m²c)	Total (kWh/m²c)
Administration	D1	92	97	189
Administration	D3	92	158	250
Administration	E1	92	115	207

2.2.3.3 Waste and Sewage Treatment

For the calculation of CO_{2eq} emissions associated with waste generation and treatment, the methodology is based on Eqs. 2.36 and 2.37, explained in the previous points (2.2.2 Tertiary: commercial and offices).

2.2.3.4 Transportation

The methodology used for the calculation of greenhouse gas in a public facilities area is similar to that used in a residential area, with two exceptions.

A first characteristic is that on this occasion the area is a centre of attraction for mobility. Therefore, instead of using the 'out' attraction matrix, the 'in' attraction matrix is used (as shown in point Tertiary: commercial and offices).

On the other hand, the number of trips generated by a residential area is related to its total plot area (0.016 trips/m² built area).[23]

$$Gt_{total}(km) = \sum_{j=1}^{179}(a_{ij} \times S_{area} \times 0.016_{trip/m2} \times r_{i-j}) \quad (2.43)$$

where Gt_{total} is the total distance travelled by the population of the study area, a_{ij} is the force of attraction between two municipalities (matrix 'in'), S_{area} is the area of the parcel under study and r_{i-j} is the distance between the origin and destination municipality (Table 2.19).

2.2.4 Industrial

As a first step in estimating the emissions generated in an area destined for industrial use, the sub-areas for industrial use must be defined and differentiated by their sub-category (as shown in Table 2.20).

---

[23] Statistical Yearbook Community of Madrid.
   https://www.madrid.org/iestadis/fijas/estructu/general/anuario/ianuserie.htm

Table 2.20 Estimated electricity and gas consumption per unit of floor area by industry type

Industry	Electricity consumption (kWh/m^2)	Gas consumption (kWh/m^2)
Food, beverage and tobacco industry	177.89	165.62
Textile and leather industry	52.48	46.14
Wood and cork industry	120.42	146.56
Printing and paper production industry	129.4	191.13
Chemical, plastic and rubber industry	596.03	762.46
Non-metallic mineral products industry	375.86	845.29
Metal products industry	204.44	126.42
Electrical, electronics and machinery industry	185.42	87.59
Furniture and other manufacturing industries	27.76	4.66
Logistics	33.9	
Other	596.03	845.29

Given the diversity of sub-areas that can be found within an urban development sector and the different demands of each of them, the tool offers the possibility of introducing up to 10 different sub-areas in each sector.

For each area the following variables must be entered: name, type of industry, plot area (m^2), buildability coefficient, landscaped open spaces (m^2), the variables buildability (m^{2}c) and occupation are calculated directly by the tool.

These entered data will be used for the estimation of emission calculations for each of the sources analysed by this tool (Fig. 2.24).

2.2.4.1 Electricity and Gas Supply

For the calculation of CO_{2eq} emissions based on the demand for electricity for air conditioning and DHW, the methodology is based on Eqs. 2.5 and 2.6, explained in the previous points (2.2.1 Household).

In the study of urban developments intended for industrial use, a distinction can be made between eleven different types of activities:

- Food, beverage and tobacco industry
- Textile and leather industry
- Wood and cork industry
- Printing and paper production industry
- Chemical, plastic and rubber industry

Description of plots for industrial use in planning according to their subcategory and surface area

Area	Type of industry	Plot area (m²)		Buildability coefficient		Buildability (m²c)		Landscaped open spaces (m²)		Occupation
Area 1	Food, beverage and tobacco industry.	15,000.00	m²	0.70	m²/m²	10,500.00	m²c	1,000.00	m²	210
Area 2	Logistics	20,000.00	m²	0.70	m²/m²	14,000.00	m²c	1,500.00	m²	280
			-		-		-		-	
			-		-		-		-	
			-		-		-		-	
			-		-		-		-	
			-		-		-		-	
			-		-		-		-	
		35,000.00	m²			24,500.00	m²c	2,500.00	m²	490

Fig. 2.24 Description of plots for industrial use in planning according to their subcategory and surface area

- Non-metallic mineral products industry
- Metal products industry
- Electrical, electronics and machinery industry
- Furniture and other manufacturing industries
- Logistics
- Other.

Buildability of complementary uses corresponds to the area destined for offices, which is the air-conditioned part. Therefore, its use is similar to the administrative type.

As in the case of residential use, the tool calculates the emissions in two ways: Option 1 (according to typology and climate zone) and option 2 (according to the declaration of the energy source) (Figs. 2.25 and 2.26)

Energy Consumption Other Uses Electricity

Option 1. According to Estimated Consumption

For the estimation of CO_{2eq} emissions from energy consumption in industrial use, the tool relies on the sources [7, 9].

$$CF_{E-T}\left(kgCO_2eq\right) = \sum (ED_i(kWh/m^2c) \times EM\ (kgCO_2eq/kWh)$$
$$+ EDg_i(kWh/m^2c) x\ EMg\ (kgCO_2eq/kWh)) \times A_i(m^2) \tag{2.44}$$

where CF_{E-T} represents the total CO_{2eq} emissions of the plots, ED_i is the electricity demand of each of the plots (as shown in Table 2.20), EM is the energy mix, EDg_i is

Option nº1 ☑

Energy demand according to typology, climate zone to comply with the Technical Building Code (CTE-HE1).

?(5.3)	Area	Type of industry	Thermal demand (kWh/m²c)		Buildability of complementary uses (m²)
	Area 1	Food, beverage and tobacco industry.	158.00	kWh/m²c	m²
	Area 2	Logistics	158.00	kWh/m²c	m²
			-		-
			-		-
			-		-
			-		-
			-		-
			-		-
			-		-
			-		-

Fig. 2.25 Example of data entry in the tool. Section: Air-conditioning and DHW (compliance with CTE-HE1) Option 1. Industrial use

Option nº2 ☑
Electricity demand according to your energy certification

?(5.4)	Area	Energy certification	Global emissions (kgCO₂ eq/m²c)		Buildability of complementary uses (m²)	
	Area 1	B	5.00	kgCO₂/m²c		m²
	Area 2	B	5.00	kgCO₂/m²c		m²
				-		-
				-		-
				-		-
				-		-
				-		-
				-		-
				-		-

Example:						
Industrial area PI-1		D	14.40	kgCO₂/m²c	1,000.00	m²

Fig. 2.26 Example of data entry in the tool. Section: Air-conditioning and DHW (compliance with CTE-HE1) option 2. Industrial use

the gas demand of each of the plots (as shown in Table 2.20), EMg is the gas energy mix and A_i is the total builtable area of each of the plots.

Option 2. According to the Declaration of the Energy Source

For the calculation of CO_{2eq} emissions based on the demand for electricity for other uses, the methodology is based on Eq. 2.10, explained in the previous points (2.2.1 Household).

2.2.4.2 Drinking Water Supply

Total emissions are divided between emissions resulting from water consumption as a consequence of the type of industry and those associated with water consumption for irrigation. For the calculation of the latter, the same methodology has been used as in the case of commercial and residential use, explained in the previous point.

$$CF_{W\,industrial}\left(kgCO_2eq\right) = [\,CF_w]_{activity}\left(kgCO_2eq\right) + [\,CF_w]_{irrigation}\left(kgCO_2eq\right) \tag{2.45}$$

where $CF_{w\,industrial}$ represents the CO_{2eq} emissions on industrial use plots associated with water consumption, $CF_{w\,acivity}$ is the total estimate of emissions associated with the type of industry and $CF_{w\,irrigation}$ is the total estimate of emissions associated with irrigation.

For the calculation of emissions according to the type of industry, the water consumption necessary for the industrial process on the one hand and the consumption associated with the use of personnel on the other hand have been analysed.

Table 2.21 Water consumption per activity in industrial use[24]

Activity	Water consumption (m^3/m^2_c)	
Food, beverage and tobacco industry		8.32
Textile and leather industry		2.66
Wood and cork industry		0.92
Printing and paper production industry		10.53
Chemical, plastic and rubber industry		44.64
Non-metallic mineral products industry		8.12
Metal products industry		4.3
Electrical, electronics and machinery industry		2.73
Furniture and other manufacturing		2.17
Logistics		0.92
Other	Maximum	44.64
	Minimum	0.92
	Average	9.38

Table 2.22 Floating population density on industrial land parcels

Industrial use	0.02	inhabitant/m^2_c

$$CF_{\text{W activity}}\left(kgCO_2eq\right) = \sum \left[CW_{\text{process}}\left(m^3\right) + CW_{\text{staff use}}\left(m^3\right)\right]$$
$$\times E_{CW}\left(kWh/m^3\right) \times EM\left(kgCO_2eq/kWh\right) \quad (2.46)$$

where $CF_{\text{w activity}}$ represents the CO_{2eq} emissions associated with the water consumption associated with the activity, CW_{process} is the total estimated water consumption associated with the industrial process, $CW_{\text{staff use}}$ is the water consumption of staff, E_{CW} is the energy per water cycle of the Community of Madrid and EM is the energy mix.

To determine the water consumption in the industrial process, the values represented in Table 2.21 have been used.

In order to relate the consumption associated with the personnel, the following relationship has been used, where the number of personnel is related to the built surface area by means of the factor as shown in Table 2.1, once the number of personnel has been obtained, it must be multiplied by the estimated individual consumption (Table 2.22).

For the calculation of CO_{2eq} emissions based on irrigation water consumption, the methodology is based on Eqs. 2.34 and 2.35, explained in the previous points (2.2.2 Tertiary: commercial and offices).

---

[24] Survey on water use in the industrial sector 2006. Retrieved from: http://www.ine.es/daco/dac o42/ambiente/aguaindu/uso_agua_indu06.pdf (Instituto Nacional de Estadística, 2006). Accessed 15/03/2024.

OPTION N1 : Estimated waste per m2 built   ☑

Area	Type of Industry	Waste generation industrial process (kg /m2c)		Total consumption (kg)	
Area 1	Food, beverage and tobacco industry.	134.87	kg /m²c	1,416,135.00	kg
Area 2	Logistics	8.64	kg /m²c	120,960.00	kg
			-		-
			-		-
			-		-
			-		-
			-		-
			-		-
			-		-

Fig. 2.27 Based on an estimated per square metre built. Industrial use

2.2.4.3 Waste and Sewage Treatment.

As in the study of areas of commercial and service uses, in this use, greenhouse gas emissions are also estimated using two options, depending on the information available on the final activity to be implemented.

Based on an Estimate Per Square Metre Built

In the first option, depending on the type of industry, the generation of waste per square metre built is estimated based on the following equation (Fig. 2.27):

$$CF_{WTG-T}(kgCO_2eq) = \sum A_i(m_c^2) \times CTG_{unitary\text{-}activity}(kg_{residue}/m_c^2)$$
$$\times EF(kgCO_2eq/kg \text{ residue}) \qquad (2.47)$$

where CF_{WTG-T} represents the total CO_{2eq} emissions associated with waste generation and treatment, A_i is the square metres built, $CTG_{unitary\text{-}activity}$ is the estimate of waste generated per industrial activity and EF is the greenhouse gas emission factor, 1.03 $kgCO_{2eq}$/kg waste

In Spain, the national statistics institute INE, provides an annual report/survey on the waste generated in the country (Table 2.23).

The greenhouse gas emission factor is taken as (1.02897 kg CO_{2eq}/kg waste).[25]

---

[25] Guía práctica para el cálculo de emisiones GEI. Oficina Catalana de cambio climático.
https://descubrelaenergia.fundaciondescubre.es/files/2013/07/Guia-practica-calcul-emisiones_rev_ES.pdf
Accessed 10/01/2024.

Table 2.23 Waste generation by industrial activity[26]

INDUSTRY	Waste generation (kg waste/m^2_c)
Food, beverage and tobacco industry	134.87
Textile and leather industry	8.64
Wood and cork industry	32.5
Printing and paper production industry	106.15
Chemical, plastic and rubber industry	238,98
Non-metallic mineral products industry	354
Metal products industry	139.77
Electrical, electronics and machinery industry	63.64
Furniture and other manufacturing industries	9.88
Logistics	8.64
Other	354

According to Waste Generation Declaration

The methodology for calculating emissions based on a declaration of waste generation is similar to that used for commercial and residential uses, explained in the previous point, as shown in Eq. 2.37.

2.2.4.4 Transportation

The methodology used for the calculation of greenhouse gas in an industrial area is similar to that used in a commercial area, with three exceptions. As in a public facilities, commercial and industrial areas can be considered as centres of attraction for mobility, hence the 'in' attraction matrix should be used for the estimation.

For industrial use, it can be used for commuting of the working staff. On this occasion, the travel estimate differs from previous areas.

The objective is to determine the number of trips per built-up area. In the case of the Community of Madrid and under normal conditions this information is based on empirical studies; frequently the generation rates used in traffic are: Commercial (tertiary): 0.04 trips/built area (15% heavy vehicles); Industrial: 0.014 trips/built area (80% heavy duty vehicles); Installations: 0.016 trips/built area (Table 2.24). Built-up area is always given in square meters.

In order to determine the kilometres travelled by heavy goods vehicles and the different emission factors depending on their typology, the calculations are based

---

[26] Survey on waste generation in the industrial sector. Retrieved from: http://www.ine.es/jaxi/menu.do?type=pcaxis&path=%2Ft26%2Fe068%2Fp01&file=inebase&L=0 (Instituto Nacional de Estadística, 2016).
Accessed 15/03/2024.

Table 2.24 Estimated trips in the industrial use

	Estimated trips
Industrial use	0.014 trips/m^{2}built area

Table 2.25 Percentage of trips in heavy and light vehicles by domains

	Light vehicles (%)	Heavy vehicles (%)
Industrial use	20.00	80.00

Table 2.26 Ratio of heavy goods vehicles according to the use of the area and the type of heavy goods vehicle in the Community of Madrid[27]

	Industrial use
Ratio of van/day per 1000 m^{2}c	4.25
Ratio of trucks/day per 1000 m^{2}c	4.70

on the following ratios determined in studies by the Community of Madrid for their estimation (Table 2.25).

2.2.5 Roads and Green Areas

The tool allows the calculation of CO_{2eq} emissions as a consequence of public lighting and irrigation of public green areas. It is reminded that the watering of private green areas in each sector is calculated in the water consumption section for each particular land use (Table 2.26).

$$CF_{GA}\left(kgCO_2eq\right) = CF_{GA\text{-street lighting}}\left(kgCO_2eq\right) + CF_{GA\text{-irrigation}}\left(kgCO_2eq\right) \tag{2.48}$$

where $CF_{GA\text{-street lighting}}$, is the emissions generated by street lighting and $CF_{GA\text{-irrigation}}$ is the sum of the emissions generated by irrigation of green areas.

2.2.5.1 Lighting on Public Roads

The tool allows the calculation of emissions from public lighting in two ways. A generic way based on the size of the population of the municipality or a detailed way where the number of luminaires and their power must be entered.

---

[27] Comunidad de Madrid [7].

Option nº1 (generic option)   ☑

| Category according to size of municipality | CATEGORIA A | | Annual per capita consumption | 81.00 | kWh/inhabitant year |

Fig. 2.28 Generic option

Table 2.27 Estimated consumption of public lighting

	Municipality size	kWh/inhabitant year
> 75,000	Category A	81
40,001 to 75,000	Category B	118
10,000 to 40,000	Category C	133
< 10,000	Category D	169

Generic Form

In this way, the size of the municipality is associated with a type of category, each of which has an associated consumption per inhabitant per year (Table 2.27).

$$CF_{\text{GA-street lighting}}\left(\text{kgCO}_2\text{eq}\right) = CA_{\text{s/category}} \times \text{PAD} \times \text{EM} \ (\text{kgCO}_2\text{eq/kWh}) \qquad (2.49)$$

where $CF_{\text{GA-street lighting}}$ represents the $CO_{2\text{eq}}$ emissions associated with street lighting, $CA_{\text{s/category}}$ consumption according to the size of the study area, PAD is the population of the development area under study (sum of the population of the development area and floating) and EM is the energy mix (Fig. 2.28).

Calculation of Total Lighting Power

Based on the actual number of luminaires and their power, the $CO_{2\text{eq}}$ emissions are determined, based on an estimated usage of 4,000 h/year.

$$CF_{\text{GA-street lighting}}\left(\text{kgCO}_2\text{eq}\right) = \sum \text{Pot}_i \times \text{luminaries}_i \times \text{hours}_{\text{annually}} \times \text{EM} \ (\text{kgCO}_2\text{eq/kWh})$$
$$(2.50)$$

where $CF_{\text{GA-street lighting}}$ represents the $CO_{2\text{eq}}$ emissions associated with street lighting, Pot_i is the unit power of each luminaire type, luminaries_i is the number of luminaires of each type, $\text{hours}_{\text{annually}}$ are the estimated operating hours and EM is the energy mix (Fig. 2.29).

If the lighting project is available, the different models of luminaires can be entered in the attached table, together with their characteristics.

Trademark	Modelo de luminaria	Luminaire model	Luminous efficiency (lm/W)	Power (W)	Operating (hours/day)		Consumption (kWh)		Quantity		% solar energy supply
			lm/W	W	4,000.00	hours/year	0.00	kWh		ud.	
			lm/W	W	4,000.00	hours/year	0.00	kWh		ud.	
			lm/W	W	4,000.00	hours/year	0.00	kWh		ud.	
			lm/W	W	4,000.00	hours/year	0.00	kWh		ud.	
			lm/W	W	4,000.00	hours/year	0.00	kWh		ud.	
			lm/W	W	4,000.00	hours/year	0.00	kWh		ud.	

Fig. 2.29 Detailed option

2.2.5.2 Irrigation in Public Green Areas

The methodology for the estimation of greenhouse gas emissions associated with the irrigation of public green spaces is the same as the process explained in Sect. 2.2.2.2 Drinking water supply.

2.2.6 Sink and Mitigation

2.2.6.1 Sink

In this case, unlike in previous sections, emissions are not caused by use or activity, but by changes in land use that impact on the absorption capacity or sink of CO_{2eq}, due to differences in vegetation cover. A place where urban planning is implemented. This implies considering the carbon footprint of changes in land use that alter its capacity to act as a sink, rather than just emissions from future uses and activities.

Therefore, the role of carbon sequestration in this respect is assessed by considering land use change from the baseline to the planned situation.

Analysing the CO_{2eq} sequestration of the vegetation cover of the area, distinguishing tree species: poplars, conifers, conifers and other tree species, irrigated arable crops, rainfed crops, shrubs, coniferous shrubs, other non-coniferous shrub species, etc. According to the study as a source, a total of the difference in CO_{2eq} capture for 28 possible coverages. Finally, public green areas and landscaped open spaces are also considered as permeable surfaces in the area.

Developed surfaces are considered impermeable, and if strategies are implemented to make them impermeable so that there is no soil sealing, mitigation measures can be considered for assessment as additional measures.

2.2.6.2 Mitigation

All measures adopted for climate change mitigation at the urban planning stage are analysed here, and therefore as conditions of the development under analysis. Self-consumption and reclaimed water measures are included which have a direct impact on the emissions derived from the energy saved.

Therefore, the overall CO_{2eq} emissions in this field are the result of applying Eq. 2.51:

$$CF_{CAP}(kgCO_2eq) = CF_{CAP\text{-}COVER}(kgCO_2eq)$$
$$+ \ CF_{CAP\text{-}SELFCONSUMPTION}(kgCO_2eq)$$
$$+ \ CF_{CAP-REGENERATED}(kgCO_2eq) \qquad (2.51)$$

where CF_{CAP} is the footprint generated by greenhouse gas capture, $CF_{CAP-COVER}$ is the footprint generated by vegetation cover, $CF_{CAP-SELF-CONSUMPTION}$ is the footprint generated by the self-consumption measures implemented and $CF_{CAP-REGENERATED}$ is the footprint generated from the emissions derived from the energy saved in the use of reclaimed water.

$$CF_{CAP\text{-}COVER}(kgCO_2eq) = \sum Capture_i(kgCO_2eq/m^2) \times SC_i)_{inicial}$$
$$- \sum (Capture_i(kgCO_2eq/m^2) \times SC_i)_{final} \qquad (2.52)$$

where $Capture_i$ is the potential CO_{2eq} capture per unit area as a function of land cover type both before and after (as shown in Table 2.28) and SC_i is the area of each or all land cover types.

For the estimation of measures adopted for climate change mitigation at the urban planning stage, self-consumption and reclaimed water measures are studied (Figs. 2.30 and 2.31).

Self-Consumption Measures

In the analysis of each of the land uses in the point of energy consumption, the tool allows to define the % of energy coming from self-consumption.

The methodology to translate this % into CO_{2eq} emissions savings is as explained in Eq. 2.53:

$$CF_{CAP\text{-}SELFCONSUMPTION}(kgCO_2eq) = \sum (DE_i(kWh/m^2c) \times S_i(m^2) \times \% \text{ selfconsumption})_i$$
$$\times M (kgCO_2eq/kWh) \qquad (2.53)$$

Reclaimed Water Measurements

In the analysis of each of the land uses at the point of water consumption, the tool allows to define the % of water coming from reclamation. The methodology to translate this % into CO_{2eq} emissions savings is as follows (Eq. 2.54):

$$CF_{CAP\text{-}regenerated}(kgCO_2eq) = \sum (S_{green\text{-}areas}(m^2) \times Q(m^3/m^2 \text{ annual})$$
$$\times E_{CW}(kWh/m^3) \times \% \text{ regeneration})_i$$
$$\times EM(kgCO_2eq/kWh) \qquad (2.54)$$

Table 2.28 Potential capture of non-developable land uses and crops.[28]

Land uses and crops	Potential capture CO_{2eq} (tCO_2/ha)
Poplar	18.66
Conifer	19.24
Coniferous and other non-coniferous species	15.87
Irrigated arable crops	36.75
Dry crops	13.45
Shrubs	4.5
Shrub with conifers	1.55
Shrub with other non-coniferous	2.38
Weeping olive hedge	6.59
Other non-coniferous	12.5
Grass	8.82
Grass shrub	5.94
Irrigated vineyard	19.11
Dry vineyard	6.26
Dry crops with other non-conifers	19.7
Irrigated fruit trees	21.92
Dry fruit trees	6.3
Horticultural crops	12.58
Vineyard with olive trees	13.19
Irrigated olive trees	20.12
Shrub with coniferous and non-coniferous trees	1.55
Conifer with eucalyptus trees	31.26
Olive trees with conifers	12.91
Grass without conifers	3.54
Shrub-grass without conifers	5.45
Dry crop shrub with conifer and non-conifer trees	21.25
Vineyard with fruit trees	6.28
Natural Grass	6.33
Eucalyptus	43.18

---

[28] Carbon Footprint and Urban Planning Incorporating Methodologies to Assess the Influence of the Urban Master Plan on the Carbon Footprint of the City (Zubelzu and Álvarez 2016).

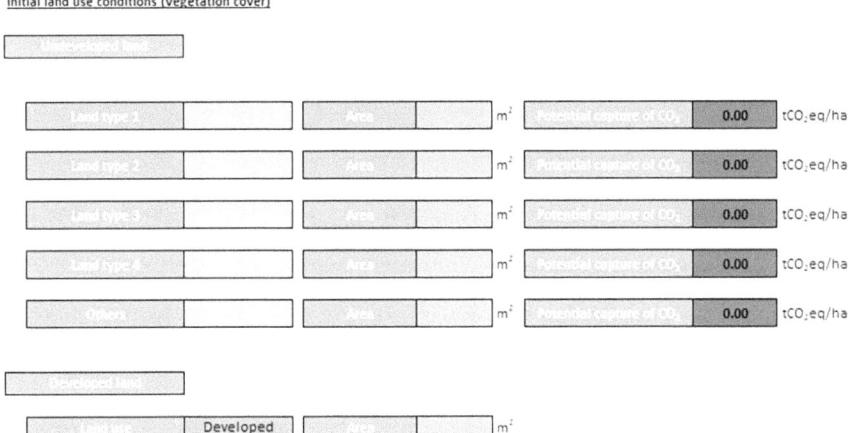

Fig. 2.30 Example of Initial land use conditions (vegetation cover)

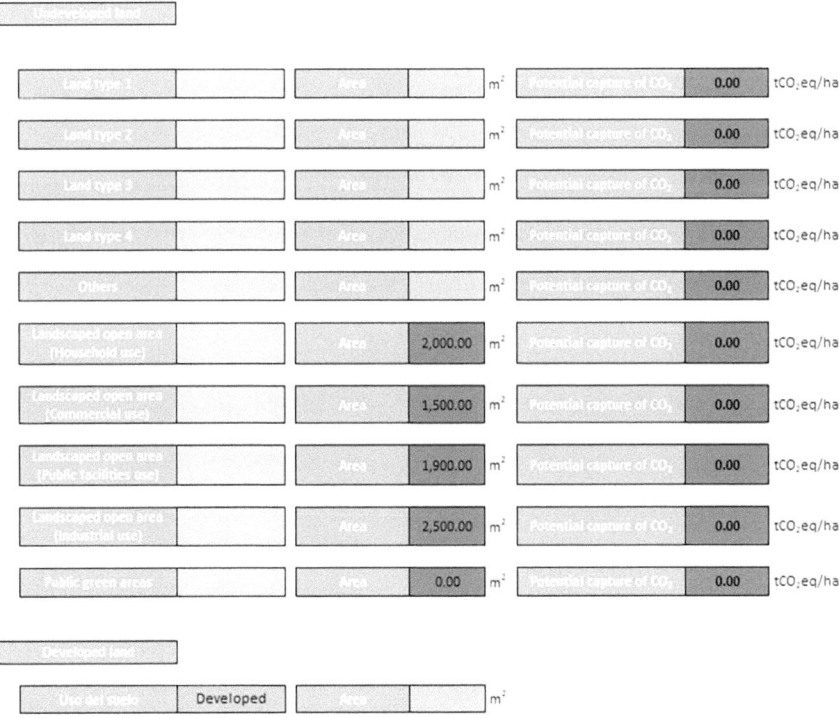

Fig. 2.31 Final land use conditions (land cover)

References

1. S. Zubelzu, R. Álvarez, A. Hernández, El papel del mix de generación de energía eléctrica y la huella de carbono en las infraestructuras urbanas, in *Proceedings of XIX International Congress on Project Management and Engineering*, 16–17 July 2015
2. R. Mertens, *Manual for Statistics on Energy Consumption in Households* (Publications Office of the European Union, Luxembourg, 2013)
3. Certificado de Eficiencia Energética, Zonas climáticas del CTE. Retrieved from https://cer tificadodeeficienciaenergetica.com/blog/wp-content/uploads/2023/01/zonas_climaticas_cte. png (n.d.). Accessed 15/03/2024.
4. Gobierno de España, *Certificación del Urbanismo Ecológico. Guía Metodológica para los Sistemas de Auditoría, Certificación o Acreditación de la Calidad y Sostenibilidad en el Medio Urbano. Agencia de Ecología Urbana de Barcelona* (Centro de Publicaciones, Ministerio de Fomento, Madrid, 2012) Retrieved from https://es.scribd.com/document/536566413/Certif-Urb-Ecosistemico-Web
5. Certificados energéticos (n.d.). Retrieved from https://www.certificadosenergeticos.com/ como-calcula-escala-calificacion-energetica
6. Oficina Catalana de Cambio Climático, Nota Informativa sobre la Metodología de Estimación del Mix Eléctrico (2020)
7. Comunidad de Madrid, Anuario Estadístico de la Comunidad de Madrid 1985–2020. Transportes y comunicaciones (2020)
8. P.A. Lopez_Jimenez, M. Perez-Sanchez, F.J. Sanchez_Romero, Huella energética del agua en función de los patrones de consumo en redes de distribución. Ingeniería del agua **21**(3), 197–212 (2017)
9. Fundación Canal, Huella energética en el ciclo integral del agua en la comunidad de Madrid (2017)
10. Comunidad de Madrid, Características de la población y los hogares a través de la Encuesta de Población Activa 2021 (2022). Retrieved from https://www.madrid.org/iestadis/fijas/efemer ides/descarga/phepa21not.pdf
11. S. Zubelzu, R.Á. Fernández, Urban planning and industry in Spain: a novel methodology for calculating industrial carbon footprints. Energy Policy **83**, 57–68 (2015)
12. Oficina Catalana de Cambio Climático, Guía práctica para el cálculo de emisiones de gases de efecto invernadero (GEI) (2013). https://descubrelaenergia.fundaciondescubre.es/files/2013/ 07/Guia-practica-calcul-emisiones_rev_ES.pdf, Accessed 15 Mar 2024
13. Comunidad de Madrid, Atlas de la movilidad residencia-trabajo en la Comunidad de Madrid (2017).

Chapter 3
Assessing the Carbon Footprint of UPMPs: A Case Study Explained in the Context of the Community of Madrid

In the following section, an example of the methodological approach for assessing the carbon footprint of an UPMP is developed, through the basis of the Community of Madrid. A set of features and municipalities as locations with different features has been selected with a view to limiting the scope, and the carbon footprint has been calculated for the urban planning major land uses: residential and tertiary.

3.1 The Community of Madrid and Its Metropolitan Region of Madrid

The Metropolitan Region of Madrid is the second metropolitan area that has seen the greatest increase in the number of inhabitants in Europe in the last five years, with a 4.9% increase of inhabitants and reaching 6.9 million people.[1] Of this population, 6.2 million are concentrated in the 52 central municipalities. These have an area of 2890 km^2 and a population density of 2165 inhabitants/km^2. Madrid accounts for 52% of this population and is the second largest municipality in Europe.

The Metropolitan Region of Madrid, as a core of the Community of Madrid, has been a hotspot of urban growth at European level with a very unique dynamic. As a result, the Community of Madrid has almost doubled the artificial surfaces from 1990 to 2018 in less than thirty years, so the processes of urban expansion have been very important for the region.

The Community of Madrid, region that includes 179 municipalities, is the only region in Spain that does not have spatial planning tools approved, i.e. a plan for the entire region or at the subregional level.

---

[1] Tamayo [1].
   Accessed on: 20/03/2024.

© The Author(s), under exclusive license to Springer Nature Switzerland AG 2024
R. Álvarez-Fernández et al., *A Carbon Footprint Calculation Tool for Urban Development*, SpringerBriefs in Applied Sciences and Technology, https://doi.org/10.1007/978-3-031-69892-7_3

This means that territorial transformations, such as the creation of a new neighbourhood or a new planning for a city or town, are decided in each masterplan of a municipality or modifications to it. Without a regional framework, although the impact in such a conurbation region is territorial, especially in topics as mobility (Fig. 3.1[2]).

Therefore, the transformation of the region has been decided on a plan-by-plan basis (sectoral, such as communications and municipal infrastructures) and not with a vision of the whole and of the interrelationships between areas, which in practice means a lack of control over the development as a whole. This also influences climate change in a specific way through induced mobility and through the occupation of the territory and the sealing of soils caused by artificialization.

Added to this is the territorial impact of the development model and the specific weight of metropolitan mobility in the Madrid region as a whole, which is much greater than in other Spanish regions. It has a very important centre-periphery mobility, commuting, which has varied through the different conformations of the metropolitan fabric, in a first wave during the developmentalist period (1940–1975) and in a second wave in recent decades (especially during the 'prodigious decade' 1997–2008) [2, 3].

This configuration and development pattern implies high levels of energy consumption and greenhouse gas emissions for the region [2, 3].

In this way, although there is still a lack of regional planning, it is possible to determine an impact that affects the whole (and globally), such as climate change. It is possible to visualize how urban transformation decisions in one area or municipality are related to the whole and can influence the urban model and, therefore, the development model. Specifically measuring the carbon footprint of an area or territory involves characterizing it in terms of its energy consumption and land use.

The need for the application of sustainable urban planning in the Community of Madrid is crucial because of the scale and population of the region and the fact that it has become a centre of accumulation and consumption [4] which, through its development model, has had an impact on climate change.

3.2 UPMP Case Study: Business-As-Usual of an Existing Neighbourhood

Sector 11B in Alcalá de Henares (Community of Madrid) is selected to show an example of visualization the impact of GHG emissions.

---

[2] Atlas de la movilidad residencia-trabajo en Comunidad de Madrid 2017.

https://www.madrid.org/iestadis/fijas/estructu/general/territorio/atlasmovilidad2017/INDEX.html

Accessed on: 12/03/2024.

Fig. 3.1 Inter-municipal flow of workers in the Community of Madrid 2016

The case study is an area developed by a masterplan of 27.27 ha located in Alcalá de Henares, a city of 193,751 of inhabitants and a surface 87.99 km². It has got about 1,250 dwellings, within a surface around half the sector. The estimated population is 3.125 inhabitants (2.5 Hs) and the estimated floating population is 905 inhabitants. The hypothesis of current situation (or alternative 0) is an undeveloped area non-agricultural grassy bush.

The main features of the sector are explained in the following data available in Table 3.1.

Table 3.1 Basic data for planning alternative 1 of the case study, Sector 11B in Alcalá de Henares, Madrid

	Use surface (m²)	Built-up surface (m²)	Occupation surface ground floor (m²)
Residential use	132.90500	110.70000	75.57150
Tertiary use	19.65000	13.75500	11.79000
Industrial use	0.00	0.00	0.00
Facilities	32.85500	30.83000	16.42750
Undeveloped	27.80000	0.00	0.00
Road system	64.007.00	0.00	0.00

This is a neighbourhood already developed, and its features have been considered. In this case with energy certification B for buildings, as a minimum for new buildings (Fig. 3.2).

This alternative presents a total of 6,774 t CO_{2eq} per year. The majority of emissions from the planned urban development by sources and uses belongs to Residential Use (53%), but with less intensity that the built-up surface that represents (71%). Mobility stands out, being the highest of all the emissions generated, because 59% of GHG emissions corresponds to the mobility generated, without considering emissions from land-use change.

In the case of this location, the average distance of trips is set at 5.28 km and the percentage of trips for the sector being 53.81% (Fig. 3.3).

The second biggest source is energy consumption (without considering air conditioning and DHW, the third source) with 19%. The land cover changes supposes the emission of 110 t CO_{2eq} because the change of an undeveloped area non-agricultural grassy bush of 27 ha ($-$ 164 t CO_{2eq}) to grass land cover of green areas, public ones (27,800 m²) and private landscaped open spaces (58,087,13 m²), with a soil´s sink capacity of $-$ 54 t CO_{2eq}.

Conclusions

Having a tool to calculate the greenhouse gas emissions of an activity or group of activities is undoubtedly great news.

The authors of this book set themselves a complex but exciting challenge: to design a tool that would allow estimating the impact on the carbon footprint applied to urban planning and the evolution of urban planning and environmental problems that have occurred to date.

Planning cities in a context of climate crisis takes on particular importance in a context of urban dynamic growth of artificial surfaces since 1990.

The carbon footprint calculator consists of an assessment of the uses and activities to be developed in future planning that generate greenhouse gas emissions, as well as changes in land use that affect sink capacity. Mitigation strategies (such as self-generation capacity through renewable energy) are analysed for evaluation and quantification where data is available.

This carbon footprint calculator includes the derived and influential activities to be requested for approval of urban planning instruments, within the ordinary or simplified procedures of the Strategic Environmental Assessment, in relation to the potential environmental impacts in terms of climate change.

The carbon footprint calculator could help to measure different urban planning alternatives. Thus, thanks to the carbon footprint calculator, it is possible to choose between the alternatives proposed those with the lowest carbon emissions among several alternatives and to make visible the crucial aspects that generate the highest number of emissions at an early stage of urban development when it comes to urban planning.

Visualization examples have been included that are completely indicative. The reader can build their own tool with the software they deem most appropriate. We hope that the approach we have given to the development of this tool will become a support for the readers when they decide to make their own calculation tool.

© The Editor(s) (if applicable) and The Author(s), under exclusive license to Springer Nature Switzerland AG 2024
R. Álvarez-Fernández et al., *A Carbon Footprint Calculation Tool for Urban Development*, SpringerBriefs in Applied Sciences and Technology, https://doi.org/10.1007/978-3-031-69892-7

References

1. Tamayo, M. (2024). Barcelona y Madrid, las áreas metropolitanas que más crecen en Europa. Eje Prime. Retrieved from: https://www.ejeprime.com/mercado/barcelona-y-madrid-las-areas-metropolitanas-que-mas-crecen-en-europa
2. A. Delgado-Jiménez, Evolución y crisis de la Región Metropolitana de Madrid 1985–2007. Análisis de las diversas perspectivas para la transformación del gobierno urbano: una revisión crítica del planeamiento. Doctoral dissertation, Arquitectura (2012)
3. A. Delgado-Jiménez, Evolución y crisis de la Región Metropolitana de Madrid: una revisión crítica del planeamiento (Editorial Publicia, 2013)
4. R. Méndez, El territorio de las nuevas economías metropolitanas. EURE (Santiago) **33**(100), 51–67 (2007)

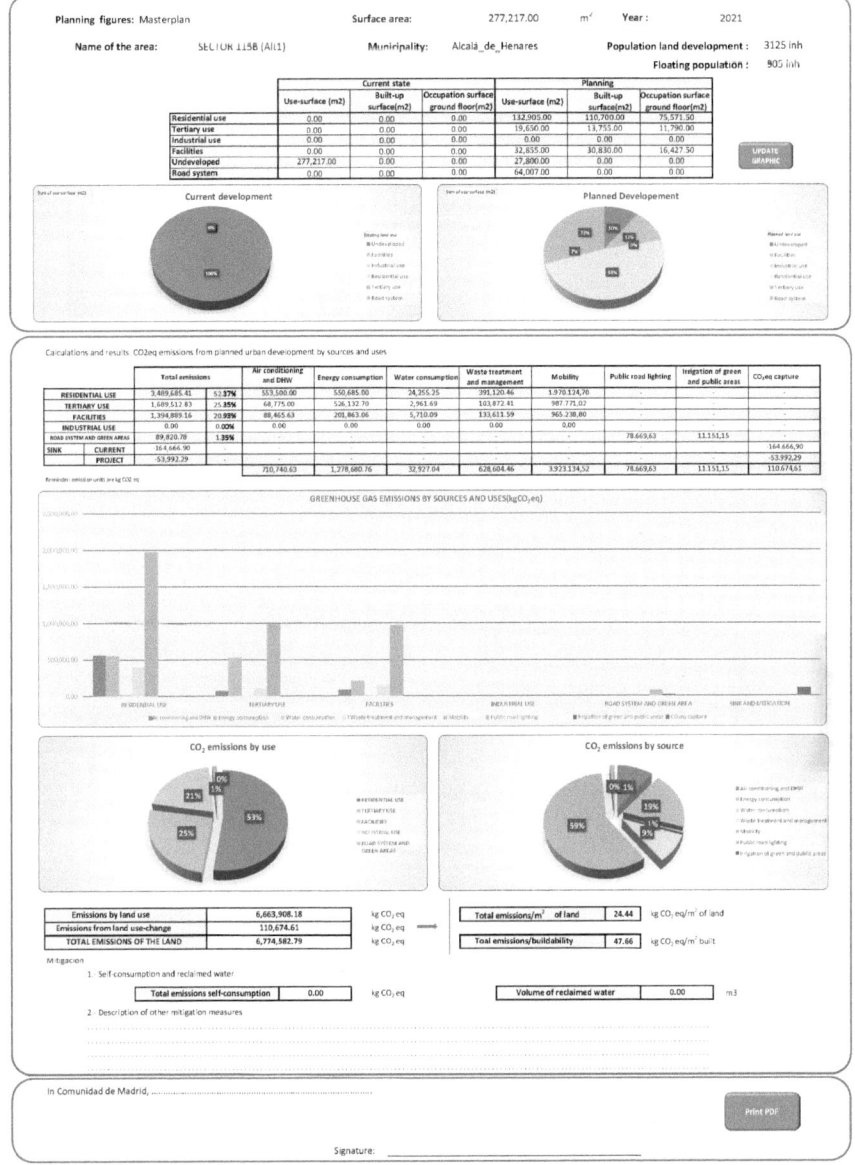

Fig. 3.3 Sector 11B Alcalá de Henares. Example of a report of carbon footprint urban planning extracted with the tool

Fig. 3.2 Community of Madrid: Municipalities for location in alternatives 1, 2 (2.1 and 2.2), 3, 4 (Alcalá de Henares), 5 (Madrid) and 6 (San Martín de Valdeiglesias) for showing the weight of location by population size and size of the municipality in the generation of GHG